Tiago Batista Cerqueira
Jamerson dos Santos
Dário Luiz Nicácio Silva

Construction and use of a Fixed Focus Parabolic Solar Disc for cooking

Tiago Batista Cerqueira
Jamerson dos Santos
Dário Luiz Nicácio Silva

Construction and use of a Fixed Focus Parabolic Solar Disc for cooking

ScienciaScripts

Imprint

Any brand names and product names mentioned in this book are subject to trademark, brand or patent protection and are trademarks or registered trademarks of their respective holders. The use of brand names, product names, common names, trade names, product descriptions etc. even without a particular marking in this work is in no way to be construed to mean that such names may be regarded as unrestricted in respect of trademark and brand protection legislation and could thus be used by anyone.

Cover image: www.ingimage.com

This book is a translation from the original published under ISBN 978-613-9-66395-8.

Publisher:
Sciencia Scripts
is a trademark of
Dodo Books Indian Ocean Ltd. and OmniScriptum S.R.L publishing group

120 High Road, East Finchley, London, N2 9ED, United Kingdom
Str. Armeneasca 28/1, office 1, Chisinau MD-2012, Republic of Moldova, Europe
Printed at: see last page
ISBN: 978-620-8-01947-1

A generation in search Neither good
nor evil The very step is the reason

New Baianos

SUMMARY

Despite technological advances and the use of more efficient energy sources for domestic cooking, firewood is still the dominant energy source in rural areas. This is even more worrying in the sertanejo region, due to the impacts such as deforestation and desertification of the caatinga and cerrado. Due to the climatic characteristics of the sertão, this region has high potential for the use of solar energy. In view of this, the Fixed Focus Parabolic Solar Concentrator could be an option for mitigating the impacts of collecting and burning firewood for cooking, by using solar energy for this purpose. This technology, developed in the 1980s by Austrian physicist Wolfgang Scheffler, consists of a point-focus solar concentrator model that has a mechanism for tracking the sun's apparent movement and a seasonal adjustment mechanism that allows the focus to remain fixed throughout the day and year. This innovative feature makes it possible for the receiver (oven) to be in the shade, for example inside a kitchen, making the use of solar energy for cooking more convenient and practical. This work describes the construction of a 2.7m Fixed Focus Parabolic Solar Concentrator^2 which was installed in the municipality of Marechal Deodoro / AL. The process of building the reflector was divided into six stages, plus the construction of a solar cooker (pot holder). To assess the practical applicability and potential of this concentrator model, five different lunch meals were prepared over five consecutive days. The temperature at the focal point was measured using an infrared thermometer over two days, where temperatures in excess of 500°C were recorded under clear skies. The results showed the potential of this concentrator for the food cooking process. Despite the benefits and applications of the Fixed Focus Parabolic Solar Concentrator, this technology is still not widely used in Brazil. It is hoped that this work will help to spread the use of solar concentrators and that it can help in the construction of other Scheffler disc prototypes.

Key words: Solar energy. Clean Technologies. Solar Concentrator. Scheffler disc.

SUMMARY

CHAPTER 1

INTRODUCTION

Awareness of environmental problems, which affect both the local and global scale, as well as humanity's ever-increasing dependence on energy, has been growing. As a result, nations have taken part in international events such as the Stockholm Conference (1972), Rio 92 and Rio+20, and have ratified documents such as the Kyoto Protocol (which came into force in 2005) with a shared commitment to reducing harmful gas emissions and environmental degradation. In view of this, the importance of and need for the development of clean technologies can be seen.

Solar technologies are therefore promising for a new, cleaner scenario. The sun is a clean, inexhaustible and free source of energy that can easily be utilised in two ways: passively or actively. Passive use is the result of architectural planning that harnesses sunlight to heat and light a room, as in a greenhouse. Active use is achieved through devices that optimise and apply solar energy for a specific purpose, such as the production of electricity by photovoltaic panels and thermal energy to generate steam, heat water and cook food.

Unfortunately, it is still a reality for Brazilian communities to live with the rudimentary practice of cooking food using firewood, which as well as being harmful to human health itself because of smoke inhalation, causes various environmental problems, such as desertification, reduced air quality, reduced biodiversity, climate change and global warming, among others.

It is believed that it is possible to improve the quality of life of Brazilian families by following the example of India, where certain rural communities organise community kitchens that use sunlight to cook food, using Fixed Focus Parabolic Solar Concentrators. Despite having high potential for harnessing solar energy, especially in the sertanejo region, many Brazilian families depend entirely on firewood collected from the caatinga, making the region more vulnerable to desertification.

The Fixed Focus Parabolic Solar Concentrator, also known as Scheffler's Heliostat in honour of its inventor, Austrian physicist Wolfgang Scheffler, is a technology that concentrates a beam of light on a certain region of space, a focal point, providing temperatures at that point as high as those obtained inside a wood-burning stove, using only solar energy.

This solar concentrator model has an ingenious structure that makes it easy to adjust the concentrator seasonally according to the changing seasons. As well as a solar tracking system that makes the reflector panel rotate around an axis parallel to the Earth's polar axis, moving it in synchronisation with the apparent movement of the Sun.

These innovative mechanisms (seasonal adjustment and daily rotation) make this prototype concentrator different from other models, as it ensures that the focal point always remains in the desired location, making it more efficient and practical to use concentrated solar energy, even allowing the focus to be on a covered location, such as inside a kitchen.

The aim of this work is to build a Fixed Focus Parabolic Solar Concentrator and a solar cooker, which were installed at IFAL (Marechal Deodoro campus) for practical cooking experiments and analyses of the temperature of the focal point of the reflector panel. To assess the practical applicability and potential of this concentrator model, meals were prepared with sunlight at lunchtime over a period of five days, and focal point temperatures were measured using an infrared thermometer over two days, in which temperatures of over 500°C were recorded.

In addition to the time spent during the initial study and construction of this concentrator, there was a financial outlay. The concentrator costs approximately 2,000 reais. The total cost of building the system can vary depending on the composition of the material used and price variations in different regions. Another factor that can make its construction less expensive is the use of recycled materials found in scrap yards, as was partly the case with this prototype.

The following chapters are distributed in the following sequence:

Bibliographical Review, which discusses energy, some energy sources distinguished between non-renewable and renewable resources, and finally delves into the literature on the Fixed Focus Parabolic Solar Concentrator; Materials and Methods, which describes in detail the methodology applied to build this concentrator; Application of the Fixed Focus Parabolic Solar Concentrator, where the practicality of using the concentrator for cooking is assessed; Results and Discussions, in which the temperature curve at the focal point of the reflector panel is analysed; and Final Considerations, where some conclusions drawn from this work are considered, and some proposals for future applications of the concentrator are made.

CHAPTER 2

LITERATURE REVIEW

This section will briefly discuss the concept of energy, relating it to humanity and the environment. It will then discuss the most common and promising energy sources, distinguishing between non-renewable and renewable resources. Finally, references will be made to the Fixed Focus Parabolic Solar Concentrator, considering its principles, functionality and applicability.

2.1 Energy and the Environment

In order to define energy, it is necessary to associate it with a specific phenomenon. The most widespread definition in various studies is that energy is related to the ability to produce work. In the sense that, when considering a system made up of one or more bodies, this system has energy if it can move or make "things" work (SAMPAIO; CALÇADA, 2005).

Energy is governed by the laws of thermodynamics and one of its principles is that it can never be created or destroyed, only converted from one form of energy to another, for example, the conversion of kinetic energy from the winds into electrical energy by generators, or solar energy into chemical energy by photosynthesis (HINRICHS; KLEINBACH, 2003).

The types of energies that exist in nature and their known forms are described in a straightforward manner by Hinrichs and Kleinbach:

> One of the basic types of energy is that associated with the movement of a body. We call this type of energy kinetic energy. A moving car or a rotating axle have kinetic energy. There is also energy associated with the position of a body, called potential energy. A stretched spring or a ball positioned on a table have potential energy. Kinetic and potential energy can be classified as forms of what we call mechanical energy. Other forms of energy are chemical energy (fossil fuels and food), nuclear energy (found inside the atomic nucleus), thermal energy (a heated body), light (or radiant) energy and electrical energy (HINRICHS; KLEINBACH,

2003, p. 32).

According to the International System of Units, the Joule (J) is the unit that represents the quantity of energy. However, this same quantity is defined in other units. Some of the main ones are described in Table 1, also showing the conversion factors.

Table 1 - Units of energy and power

1 joule (J)	10^7 ergs
1 lime	4,18 J
1 BTU - British Thermal Unit	252 cal
1 Kcal	1,000cal
1 W/h	3.600 J

Source: Adapted from GOLDEMBERG; VILLANUEVA; DONDERA (2003, p. 43).

The study of human development is directly related to the amount of energy consumed. Studying this relationship, Goldemberg and Villanueva categorise the development achieved by mankind into six stages:

- Primitive man (East Africa, approximately 1,000,000 years ago) without the use of fire had only the energy of the food he ate (2,000 kcal/day).
- The hunter (Europe, around 100,000 years ago) had more food and also burned wood for heat and cooking.
- Early agricultural man (Mesopotamia in 5000 BC) sowed crops and used animal energy.
- Advanced agricultural man (north-western Europe in 1400 AD) used coal for heating, water power, wind power and animal transport.
- Industrial man (in England in 1875) had the steam engine.
- Technological man (in the USA in 1970) consumed 230,000 kcal/day (GOLDEMBERG; VILLANUEVA, 1998, p. 29-30).

Today, humanity is capable of harnessing energy in all its different forms. However, for a long time in history, it only used the energy resulting from muscular strength, which was used simply to obtain the food necessary for survival (REIS et al, 2005).

The use of mechanical energy has been of great importance to the development of humanity, where at first it was applied through the use of animals for transport and farming, moving on to the use of instruments that allowed the application of the kinetic energy of the wind to carry out activities (REIS et al, 2005).

The intensification of industrial, agricultural and commercial activities, caused by the advance in the use of mechanical energy, led to a great demand for resources, prompting man to use other sources of energy, thus initiating the exploitation of fossil resources. According to Reis (2005), the intensified exploitation of fossil resources for energy generation began with the widespread use of mineral coal.

Nowadays, in order to satisfy his basic needs, obtain comfort and leisure, man consumes a large amount of energy, which society is increasingly dependent on, as Ornellas observes:

> In order to live today, people use diesel oil, paraffin, petrol, natural gas and coal, which are mainly used for transport, home electrification and means of production in general. In today's society, especially in the first world, almost everything man uses is industrialised, from the water he drinks to the food he eats, to the environmental climate in which he lives, moves and works (ORNELLAS, 2006, p. 61).

The current production model, based on the exhaustive exploitation of natural resources to achieve economic development, has resulted in considerable risks of negative impacts on the environment, such as air pollution and soil and water contamination. As Marcatto emphasises:

> The current model of development, which is unequal, exclusionary and depletes natural resources, has led to alarming levels of soil, air and water pollution, the destruction of animal and plant biodiversity and the rapid depletion of mineral reserves and other non-renewable resources in practically every region of the world. These degradation processes have their origins in a complex and predatory model of exploitation and use of available resources [...] (MARCATTO, 2002, p.8).

Paying attention to these issues, in the last decades of the 20th century, environmental movements in various parts of the world sparked greater interest in the relationship between economic development and environmental preservation. As can be seen below:

> Environmental issues made headway on the world stage in the aftermath of the Second World War, more precisely in the 1960s, when ecological discourse began to move from the realm of reflection to more consistent interaction, opening the way for practical environmental actions. Over the course of the last four decades of the 20th century, publications and events began to appear in various parts of the world with a vision that was more concerned with the defence of the environment (BARROS FILHO; CRUZ, 2011, p.28).

Throughout the period from the 1960s to the present day, various oppositions have arisen to the impacts of economic development on the environment. It began in 1962 with the publication of Raquel Carson's book "Silent Spring", which criticised the use of pesticides and their ecological effects. Among other environmental events on the planet, it is worth highlighting the United Nations Conference on the Human Environment (UNCED) in 1972 in Stockholm, Sweden, where the United Nations Environment Programme (UNEP) was created, and the United Nations Conference on Environment and Development (UNCED) in 1992 in Rio de Janeiro, Brazil. These conferences resulted in important documents that serve as a treaty committing many countries to environmental issues, such as Agenda 21 and the Kyoto Protocol, which, among other objectives, aims to reduce greenhouse gas emissions (BARROS FILHO; CRUZ, 2011).

2.2 Non-renewable energy sources

Every energy resource is classified as either a renewable resource or a non-renewable resource. According to Bauman (2008), a resource is called renewable or non-renewable according to the time it takes to be replenished in nature, taking into account the scale of human life. It is therefore understood that non-renewable sources take a long time to regenerate, while renewable sources are replenished in a short period of time.

According to Luíz (1985), the main non-renewable energy sources used by humans come from fossil fuels, which are substances of mineral origin, produced by the fossilisation of carbohydrates that were mostly produced by the plant kingdom.

In the energy context, fossil fuels are of significant importance. Currently, according to Bauman (2008), approximately 80 per cent of the energy used in the world comes from fossil fuels.

Beneduce (2000) describes the main reasons why fossil fuels are the main and most important energy base for today's society: they are easy to store; they have a high energy value; they have deposits in various parts of the world; and they are cost-effective.

However, some negative factors should be noted when dealing with fossil fuels.

Beneduce (2000) reports that, despite being considered relatively low-cost, they can become more expensive when deposits are exploited in remote locations. Furthermore, the worst factors linked to the use of fossil fuels are that, as well as being limited resources in nature, they are highly polluting.

The use of fossil resources has played a major role in the development of humanity. However, the intensification of their use has had harmful consequences for the environment, such as "emissions of local pollutants, greenhouse gases and jeopardising the planet's long-term supply" (GOLDEMBERG; LUCON, 2007, p. 1).

Environmental impacts such as global warming and acid rain are associated with the release of harmful gases into the atmosphere from the burning of fossil fuels. According to Goldemberg and Villanueva (2003), the five main air polluting gases are: sulphur oxides, nitrogen oxides, carbon monoxide, particulate matter, ozone and hydrocarbons.

With regard to fuel consumption (non-renewable and renewable), Achão (2003) notes that the residential sector accounts for 35.8 per cent of fuel consumption, used mainly for cooking food and heating water. Various types of fuel are used for cooking, "such as solid (charcoal, firewood, coal), liquid (paraffin), gaseous (LPG, biogas) and electricity". (SANGA, 2004.p.75). The application of biomass in primary energy consumption is used on a larger scale (80% to 90%) in countries with low purchasing power. The main fossil fuels used for cooking are paraffin and liquefied petroleum gas (LPG) (SANGA, 2004).

In 2000, world demand for LPG was approximately 200 million tonnes, with an average consumption growth rate of around 5% per year. However, consumption of this fuel is very limited in rural areas due to the lack of infrastructure for commercialisation (SANGA, 2004).

In the conventional process of cooking by burning solid fuels, a lot of damage is caused to both human health and the environment. Sanga (2004) highlights the main problems associated with this practice, such as indoor air pollution, deforestation and vulnerability to climate change. These causes are observed mainly in developing countries because they make greater use of biomass as a fuel for residential cooking purposes.

According to Ugucione (2009) one of the main gases harmful to human health released by burning LPG during food preparation is nitrogen oxides, which can damage lung function. However, not only because of better energy efficiency, "liquid and gaseous fuels are relatively cleaner than solid fuels because they emit smaller quantities of CO_2 and products of incomplete combustion, such as CO and CH4" (SANGA, 2004.p.76).

2.2.1 Mineral coal

Coal was formed millions of years ago from the decomposition of plant material which, when "[...] subjected to high pressures and temperatures in contact with air, is transformed into a solid, dark-coloured product whose physical and chemical properties depend on its geological formation" (REIS et al 2005, p. 212).

In discussing coal consumption, Hinrichs and Kleinbach (2003) report a drastic change in the market, where the use of coal has become negligible for locomotives that are now powered by electricity and diesel, but which in the 1940s consumed 125 million tonnes of coal annually. In recent years, the only sector to increase coal consumption has been thermoelectric power stations, which in 2003 consumed approximately 90 per cent of the total coal produced.

Despite the low calorific value of coal compared to oil, its exploitation is still quite significant in the world, as reported by Reis (2005):

> Coal is still a highly exploited fossil resource worldwide. It is an important vector in the economic development of many countries. Although it has been replaced by oil, nuclear energy and natural gas in some applications, it currently occupies second place in the world's energy matrix, behind oil, and is the most abundant resource in nature (REIS et al, 2005, p. 214).

Coal consumption in Brazil is of little importance, accounting for only 0.5 per cent of world consumption and only 7 per cent of the national energy matrix (REIS et al 2005).

Among the main fuels used to generate energy, coal is one of those that contributes to environmental degradation, in terms of the release of greenhouse gases. However, according to Reis et al (2005), in recent years, with increased concern for environmental quality, the consumption of this fossil fuel has been resisted by environmental organisations.

2.2.2 Oil

Like other fossil resources, oil is the result of the decomposition of organic matter over thousands of years. Its composition is basically made up of carbon and hydrogen, in the form of hydrocarbons (REIS, 2005).

Currently, oil plays a significant role in the world's energy matrix, mainly in the transport sector through its derivatives such as petrol, paraffin and diesel. This resource is also a fundamental raw material for the formation of various by-products, such as lubricating oils, plastics, paraffin, among others (MARIANO, 2001).

In order to be converted into useful products, oil undergoes a refining process which is followed by a series of stages. The main oil derivatives and their uses are described below by Reis et al (2005):

- Liquefied petroleum gas (LPG) - consists of propane and butane, or a mixture of these hydrocarbons. Obtained from natural gas or by refining crude oil. It is mainly used in the residential sector.
- Petrol - a volatile, flammable liquid that is an extremely complex mixture of hydrocarbons.
- Paraffin - Intermediate between petrol and diesel, it is obtained by fractional distillation of crude oil. It is used as fuel for jet aircraft turbines.
- Diesel oil - Fuel used in engines that operate according to the diesel cycle.
- Fuel oil - A product that is primarily burnt to produce heat.
- Lubricant - There are hundreds of lubricants made from petroleum, each with a specific purpose.
- Paraffin - refers to a versatile commercial product with a wide range of industrial applications.
- Asphalts - Dark-coloured binding materials made up of complex mixtures of non-volatile hydrocarbons with a high molecular mass (REIS et al 2005, p.187-188).

With regard to the distribution of the use of oil derivatives in the world, Ornellas (2006) noted that in 2002 oil was consumed in the following fractions: 10.3% as fuel oil; 13.4% as petrol; 36.7% as diesel; 9.8% as naphtha; 7.9% as LPG and 22.8% as other derivatives.

With the development of technologies and the increase in the rate of consumption in recent decades, the rate of total average energy consumption by the population has also increased, revealing a widespread dependence on oil and its derivatives. Goldemberg and Lucon emphasise the daily consumption of energy per person:

> In 2003, when the world population was 6.27 billion, the average total energy consumption was 1.69 tonnes of oil equivalent (toe) per capita. One tonne of oil is equivalent to 10 million kilocalories (kcal), and the average daily energy consumption is 46,300 kcal per person (GOLDEMBERG; LUCON, 2007, p. 1).

This factor poses a potential risk to the environment, as the burning and production of oil derivatives are the main emitters of gases responsible for acid rain and global warming, such as carbon monoxide and dioxide, sulphur dioxide, soot, among others.

Ornellas (2006) cites another problem in dealing with oil, which is the danger of accidents during its extraction and transport, which when they occur contaminate large areas, damaging ecosystems, as happened with the oil spill at the end of 2011 in the Campos Basin in Rio de Janeiro. The importance of looking for ways to meet the energy needs of oil through other sources that are less harmful to the environment is therefore clear.

2.2.3 Natural gas

"Natural gas is made up of a mixture of hydrocarbons, with methane (CH_4) being the main constituent of the mixture" (LUIZ, 1985, p. 5). According to Hinrichs and Kleinbach (2003), natural gas can most often be found accompanied by oil, thus called associated gas, or it can be found in the absence of oil in a reservoir, called non-associated gas.

At the beginning of oil extraction, the energy potential of natural gas was unknown and it was simply burnt off during extraction, as Luis (1985, p. 5) reports: "In the first oil explorations, natural gas was simply burnt off at the top of the well. Subsequently, it was realised that this gas had a high calorific value and that it would therefore be advantageous to exploit it".

According to Reis et al (2005), the use of natural gas in cold climate countries is mainly for heating rooms, while in Brazil it is used for cooking food. In commerce and services (bars, hotels, shopping centres, etc.), it is used as a substitute for LPG (liquefied petroleum gas). In the automotive sector, it is used in cars as a complement or substitute for petrol, alcohol and diesel oil.

In terms of the environment, natural gas is less harmful to the environment than oil and coal in terms of the release of gases during combustion. However, as it is a fossil fuel, it does have its share of environmentally harmful gases. "The main components found in natural gas are hydrogen and water vapour and its main contaminants are carbon dioxide and hydrogen sulphide gas" (REIS et al, 2005, p. 194).

2.3 Renewable Energy Sources

Unlike non-renewable energy sources, which take millions of years to regenerate in nature, renewable sources are those that can be replenished in a matter of years or less, in proportion to the human scale, as is the case with biomass. Also considered renewable are energy sources whose use by humanity does not considerably affect their potential, such as solar energy and the kinetic energy of wind and water. Some of these renewable sources are briefly described below.

2.3.1 Biomass

From an energy point of view, biomass is any renewable resource that comes from organic matter (animal and/or vegetable), such as firewood, animal waste, agricultural by-products and charcoal. In this work, more emphasis will be placed on firewood, as it is the oldest energy source used by humans, who burned it mainly for lighting and cooking food.

One of the points to consider about firewood is its low energy efficiency, which converts only 10 per cent of its contained energy into useful energy, as well as the fact that its constant use causes serious damage to the environment and human health (GOLDEMBERG; VILLANUEVA, 2003).

Despite technological advances and the use of more efficient energy sources for domestic cooking, "[...] firewood is in fact the dominant energy source in rural areas [...]" (GOLDEMBERG; VILLANUEVA, 2003, p. 56). To give you an idea, calculations show that in 1991 residential firewood consumption in Brazil was 28,462 x 10^3 tonnes, concentrated in the Northeast region with approximately 39% of this consumption (ACHÂO, 2003).

More recently, two case studies - one carried out in Minas Gerais and the other in Goiás - found in common that the consumption of firewood in a household with around four residents is close to 10kg/household/day (LÓPEZ; SILVA; SOUZA, 2000), (VALE; RESENDE, 2003).

Firewood consumption is linked to parameters such as purchasing power and fuel accessibility. Given the predominance of remote communities in the sertanejo region, biomes such as the cerrado and caatinga, which are unique to Brazil, are being jeopardised by deforestation resulting from the purchase of firewood (ALMEIDA; et al, 2008).

Environmental impacts such as "[...] deforestation and soil degradation are due, in part, to the use of firewood for cooking" (GOLDEMBERG; VILLANUEVA, 2003,

p. 73). These factors have repercussions on environmental problems that range from the local/regional scale, such as soil erosion, desertification, a decrease in air quality and a reduction in biodiversity, to the global scale, such as climate change and global warming.

Burning wood to prepare food is also one of the main causes of respiratory problems in the world, due to the high concentration of smoke and soot in the air, especially indoors. As Goldemberg and Villanueva (2003, p. 80) point out, the "[...] levels of pollution produced in homes and kitchens represent a dosage of particulates equivalent to smoking several packets of cigarettes a day", in other words, it is highly harmful to health.

Therefore, although this type of fuel is low-cost for certain rural communities, methods should be sought to reduce its consumption, such as improving the accessibility of more efficient fuels and/or alternative technologies for this domestic purpose.

2.3.2 Hydroelectric

The production of energy from hydroelectric power stations is based on converting the hydraulic energy of a given water resource into mechanical energy, which is then converted back into electrical energy by a generator. Ribeiro and Bassani clearly and succinctly describe how this happens:

> **"[...] The force of the water causes the turbine to turn, transforming the** potential **energy** that exists between the level of the reservoir before the dam and the level of the river after the dam into kinetic energy. This turbine is connected via a shaft to a generator, which also moves; in the generator, the kinetic (or mechanical) energy is converted into **electrical energy" (2011, p. 7).**

In Brazil, due to its hydrographic potential, hydroelectric power stations have a superior position in electricity production. According to Ribeiro and Bassani (2011, p. 6) "[...] hydroelectric power contributes 14 per cent of the national energy matrix and almost 77 per cent of all electricity generated in the country".

In part, this statistic is relatively good, because compared to other power plants such as thermoelectric and nuclear, hydroelectric plants are safer, cause less damage to the environment and are a renewable resource (RIBEIRO; BASSANI, 2011).

However, the installation and operation of a hydroelectric dam has a negative impact on the environment, especially at a local/regional level, such as changes to the microclimate; loss of biodiversity; modification of the river's hydrology, with flooding upstream and reduced flow downstream of the dam. In many cases, the riverside population has to be relocated, which also has a major social impact (GOLDEMBERG; VILLANUEVA, 2003). Therefore, in terms of clean technologies, alternative sources such as kinetic wind and solar energy are showing promise in the energy sector.

2.3.3 Wind Energy

Electricity from a wind turbine consists of converting the kinetic energy of the wind into mechanical energy through the rotation of the rotor's propellers, which is then converted into electrical energy by a generator (MARTINS; GUARNIERI; PEREIRA, 2008).

Since ancient times, man has used mechanical energy from the kinetic energy of the air in windmills, mainly for pumping water and grinding grain. However, it wasn't until the end of the 19th century, with the expansion of the use of electricity, that projects began to be developed to produce electricity using aero generators (DUTRA, 2001).

Currently, "wind energy has been identified as the most promising renewable energy source for electricity production" (MARTINS, GUARNIERI; PEREIRA, 2008, p. 2). Among renewable sources, wind power plants have seen surprising growth in global electricity production in recent years, to the point where, if the factor persists, it is estimated that by 2020 wind power will reach a 10% share of total electricity production and up to 20% by 2040 (DUTRA, 2001).

As with any energy park, wind farms also have negative impacts on the environment, the most notable of which are noise pollution due to the rotation of the propellers, visual impact and damage to fauna, especially birds, which run the risk of colliding with the propellers. However, with proper planning, these impacts can be minimised and/or even eliminated.

However, the impacts caused by a wind farm are relatively negligible when compared to the environmental damage caused by a thermoelectric or hydroelectric plant, because not only is wind - the driving force - a free and renewable resource, but the operation of a wind farm does not emit CO_2 or any other type of gas that contributes to the greenhouse effect, global warming and acid rain, and 99 per cent of the area of the wind farm can be used for other purposes, such as agriculture and/or farming (TERCIOTE, 2002).

Thus, for large-scale electricity production, among renewable sources, wind technologies have enormous potential for sustainability in the energy sector. It is therefore essential that investments and public policies are geared towards this sector.

2.3.4 Solar Energy

The sun is the oldest source of energy available on earth and one of the determining factors for the existence of life on the planet. Solar energy is directly and indirectly associated with all the other forms of energy present in nature and enjoyed by man, such as hydraulic energy from the synergy of wind and water resources; the chemical energy of photosynthesis produced by plants, which represent both the base of the food chain, the production of biomass, and the principle of fossil fuel formation (LUIZ, 1985; GRUPO 07, 1979).

Solar energy comes from constant and turbulent nuclear reactions in the Sun's core, in which lighter elements such as Hydrogen and Helium fuse to form more complex elements such as Oxygen, Carbon, etc. Although this energy produced in the core reaches the surface of the star merely by convection, it is propagated through electromagnetic waves through space (DELERUE, 2002).

On clear days and at certain times of the day, the maximum irradiance reaching the Earth's surface is 1000 W/m^2 . It is estimated that "the solar energy that reaches us is 80% visible light, 4% ultraviolet rays and 16% infrared rays" (CAMPBELL, 1978, p. 36). Figure 1 shows the proportions of electromagnetic radiation emitted by the sun. According to Kreith (1977), the band of the spectrogram that corresponds to this energy is also understood as thermal radiation, i.e. this energy is available in the form of heat (Figure 2).

However, despite technological advances to harness all this energy, it is estimated that small-scale projects are more viable, as Luiz (1985) notes below:

> Due to the diffuse characteristics of solar energy, it should be used locally to supply energy to homes or small communities. We believe that large-scale projects are not advantageous for harnessing solar energy (LUIZ, 1985, p.19).

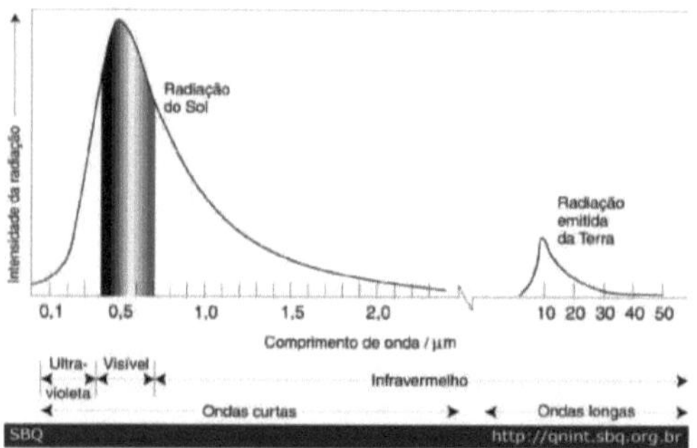

Figure 1- Solar Electromagnetic Spectrum

Source: TOLENTINO; ROCHA FILHO, 1998, p.12.

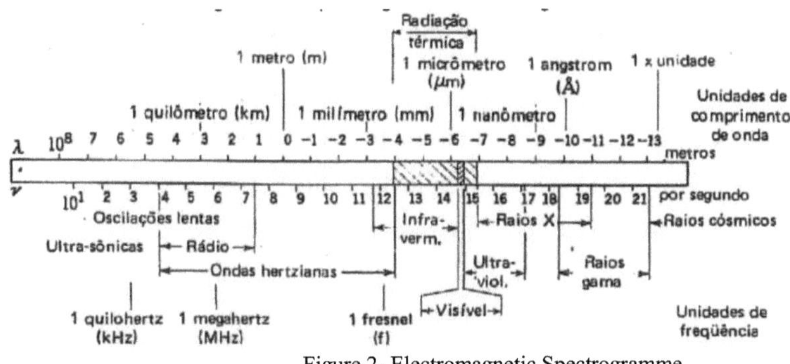

Figure 2- Electromagnetic Spectrogramme
Source: KREITH, 1977.

However, in addition to these characteristics, which are in line with the absorbance and reflectance of the atmosphere and the earth, climatic factors such as cloudiness and rainy periods, and especially nightfall, hinder the feasibility and applicability of solar energy in large-scale projects.

Regardless of the proportion, solar energy can basically be harnessed in two ways. One is through passive use, in which through architectural planning, solar energy is properly applied to illuminate and/or heat a particular environment, either directly, through glazing that transmits solar incidence to the environment - similar to a greenhouse - or indirectly, through walls that absorb and conserve thermal energy and then direct it in a desirable way to the desired enclosure (BARBOSA, 2008), (WACHBERGER, M.; WACHBERGER H., 2009).

The other way of harnessing celestial energy is through active use. Unlike passive use, this requires the use of a device that optimises and applies the energy potential to some process, such as photovoltaic cells, solar collectors and solar concentrators.

Of the solar technologies mentioned, with the exception of photovoltaic panels, which convert solar energy directly into electricity, collectors and concentrators are designed to maximise the thermal energy provided by solar incidence.

In the case of solar collectors, solar radiation is absorbed and accumulated by the panels, usually to heat water, making it reach "relatively low temperatures (below

100°C)" (ANEEL, 2005, p. 33). The use of this technology predominates in the residential and commercial sectors.

On the other hand, solar concentrators that reach higher temperatures, exceeding 100°C, open up the possibility of being used even in the industrial sector. The National Electric Energy Agency describes this technology well.

> [...] the purpose is to capture the solar energy incident on a relatively large area and concentrate it on a much smaller area, so that the temperature of the latter increases substantially. The reflecting surface (mirror) of concentrators is parabolic or spherical in shape, so that the sun's rays are reflected onto a much smaller surface, called the focus, where the material to be heated is located. (ANEEL, 2005, p. 36).

It can be seen, therefore, that the solar concentrator is comparatively more powerful due to the high temperatures it provides and, as a result, has a greater possibility of applying the concentrated energy. However, the subject of this work concerns the construction of a solar concentrator model, called a Scheffler disc, which will be the focus of the next subchapter.

2.4 Fixed Focus Parabolic Solar Concentrator

The Scheffler Reflector or Scheffler Oven is a solar concentrator model with an innovative design, developed in 1986 by the Austrian physicist Wolfgang Scheffler and used mainly for the cooking process. It was developed in such a way that its construction was as simple as possible, so that it could be reproduced in any welding workshop with locally available materials (SCHEFFLER, 2006a).

Its geometry, apparently elliptical, is based on the lateral segment of a paraboloid, i.e. a parabola in revolution, cut by a plane parallel to the zenith, which represents a beam of light, as shown in Figure 3. In this case, sunlight is reflected laterally towards the focus, and the reflector faces the Equator (SCHEFFLER, 2006a).

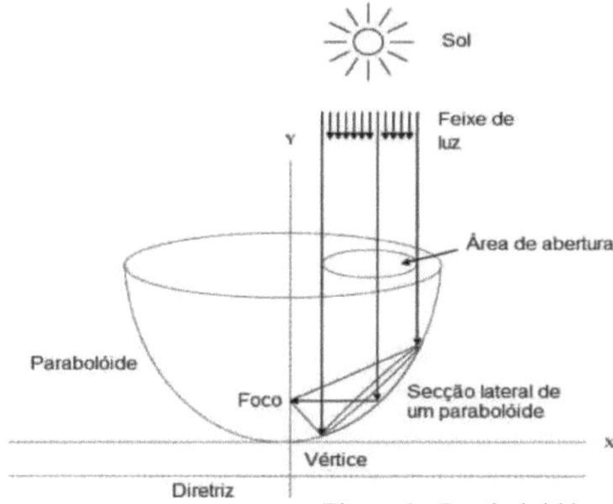

Figure 3 - Paraboloid base for the Scheffler Reflector

Source: Available at: <http://www.sciencedirect.com/science/article/pii/S0038092X10002021>.

Its operation is based on an axis parallel to the X axis and the Earth's polar axis, which simultaneously crosses the focus of the paraboloid and the centre of the reflector panel. Thus, the working axis of the reflector has a declivity identical to the latitudinal angle of the location where it will be installed. One of the novelties of this model is that it rotates around its reference axis at an angular speed of one revolution per day, i.e. approximately 15° per hour, i.e. at the same speed as the Earth's rotation, keeping the focus permanently fixed. As the axis passes through the reflector's centre of gravity, it doesn't take much force to move it (SCHEFFLER, c).

This rotational movement of the reflector is usually driven by a motor clock (equatorial tracker), fuelled by mechanical energy from a pendulum system, or by electrical energy from photovoltaic panels. The type of tracker used depends on the availability and affordability of the materials (TERRA FUNDACIÓN, 2006).

The other outstanding feature of this concentrator is that its surface is flexible enough to adapt to each change of season (seasonal adjustment), ensuring that the focus remains fixed all year round. This seasonal adjustment of the reflector is easily done by the operator using two levers. During each seasonal adjustment, both the inclination and the shape of the reflector panel (parabola formula) change. This is due to the approximately 23° variation in solar radiation in relation to the Equator at each solstice

(Figure 4) (SCHEFFLER, 2006a).

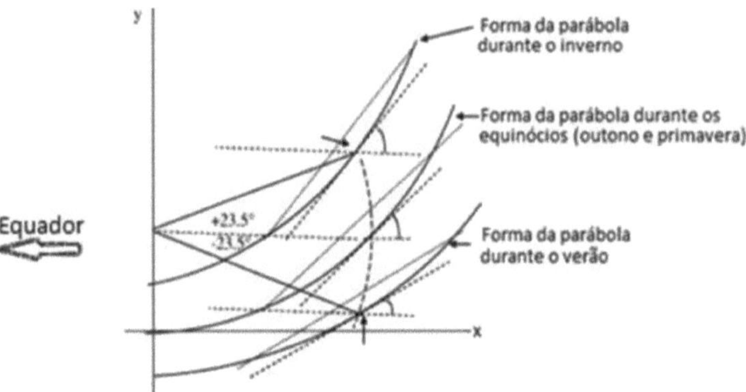

Figure 4 - Seasonal adjustment of the parabolic dish
Source: <http://www.sciencedirect.com/science/article/pii/S0038092X10002021>

These sophisticated mechanisms - daily rotation and seasonal adjustment - are what set it apart from other concentrator models and ensure that the focal point always remains in the desired location, making it more efficient and practical to apply the concentrated energy, even allowing the focus to be in a covered location, such as inside a kitchen.

It is not known how many of these concentrators are in use today, but according to Passos et al apud Hoedt (2007) in 2006 there were approximately 950 reflectors around the world, but it is estimated that more than 2000 units have been in operation recently. This technology is present mainly in Africa and Asia, and especially in India, where Scheffler Concentrators have been widely used to generate steam in community kitchens, such as a Yoga centre in Abu Road, Rajasthan and the Tirumala temple in the southern Indian city of Tirupathithat (DELANEY, 2009).

There are a wide variety of applications for concentrated solar energy. A good example is the solar crematorium project developed by Wolfgang Scheffler in the city of Valsad in Gujarat, India. In his project, Scheffler used a reflector with an effective area of 3.4 metres^2 to concentrate sunlight in a crematorium specially built for this concentrator. Inside the solar crematorium, the temperature reached 900°C, causing the total combustion of 4kg of goat meat in just 35 minutes, proving the effectiveness of this concentrator as a heat source for a crematorium (SCHEFFLER, 2006b).

CHAPTER 3

MATERIALS AND METHODS

This article describes the materials and methodology used to build a Fixed Focus Parabolic Solar Concentrator (Scheffler) and a solar cooker at the Federal Institute of Alagoas, Marechal Deodoro campus, assessing their potential and efficiency for cooking food.

The municipality of Marechal Deodoro is located in the Macroregion of Alagoas, specifically at Latitude 09° 42' 37" South and Longitude 35° 53' 42" West, at an altitude of 31 metres, the region has a hot and humid tropical climate (ANUÁRIO ESTATÍSTICO DO ESTADO DE ALAGOAS, 2010/2011). As such, the city has great potential for harnessing solar energy. Figure 5 shows the exact location of the concentrator within the IFAL Marechal Deodoro campus.

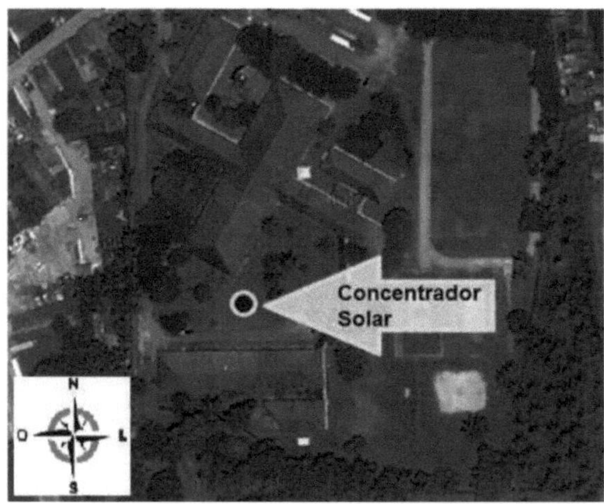

Figure 5 - Location of the solar concentrator
Source: Google Earth, 2013.

3.1 Construction of the Parabolic Focus Solar Concentrator Fixed

The concentrator was basically built in six stages: the base, the rotation support template, the rotation support, the reflector panel and the whole assembly. Next, a

26

simple cooker was developed for the direct use of solar energy. Finally, tests were carried out to assess the temperature and efficiency of cooking food.

On the last pages of this work there are annexes that helped in some stages of the construction of the concentrator, mainly in the assembly of the rotation support template and the rotation support. They consist of plans drawn up for the project by Manu Gomez and Maren Kern (2010). The work of Dib (2009) was another resource consulted for the construction of this prototype.

3.1.1 Base

The purpose of the concentrator base is to lift and support the entire solar concentrator assembly and ensure that the axis of the rotation support remains parallel to the Earth's polar axis.

The base of the concentrator is made up of 50 by 50 mm square sections of galvanised metallic steel and other metal components of unknown composition, purchased from scrap yards. Table 2 describes the parts used in the base with their respective dimensions, Figure 6 shows the drawing of the base and the nomenclature of the parts.

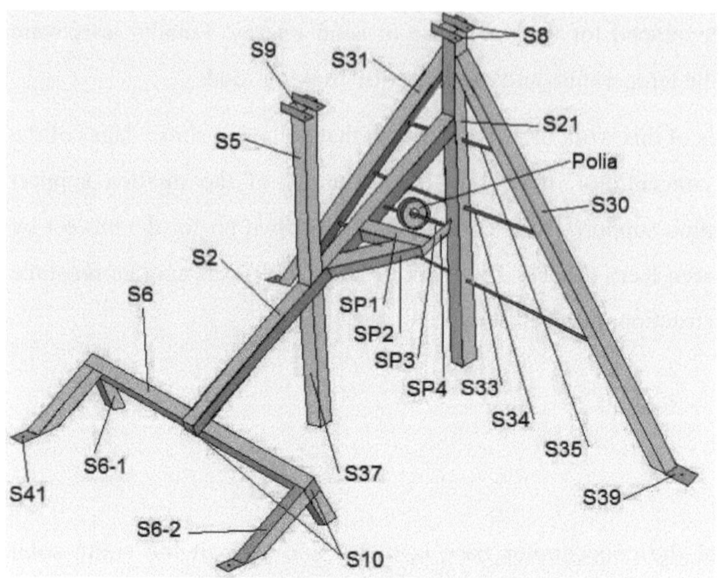

Figure 6 - Drawing of the base with the nomenclature of its parts
Source: CERQUEIRA; SANTOS, 2013.

To make it easier to build the base, it was divided into five stages, as follows: making up the back of the base; making up the front; interlocking the ends, i.e. the middle of the base; building the pulley support; and welding the "hole plug" parts.

Table 2 - List of base parts

Nomenclature	Quantity	Profile	Dimensions (mm)	Length (mm)
S2	1	Square tube	50x50x2,5	1280
S5	1	Square tube	50x50x2,5	619
S6	1	Square tube	50 x 50 x 2.5	1050
S6-1	2	Square tube	50x50x2,5	215
S6-2	2	Square tube	50 x 50 x 2.5	363
S8	4	Flat bar	25x6	75
S9	2	Flat bar	50x6	100
S10	5	Flat bar	50x6	50
S21	1	Square tube	50x50x2,5	1344
S30	1	Square tube	50x50x2,5	1471
S31	1	Square tube	50 x 50 x 2.5	1471
S33	2	Circular bar	10	640
S34	2	Circular bar	10	450
S35	2	Circular bar	10	200
S37	1	Square tube	50 x 50 x 2.5	583
S39	2	Flat bar	50x6	115
S40	2	Flat bar	50x6	60
S41	2	Flat bar	50x6	150
SP1	1	Square tube	50 x 50 x 2.5	180
SP2	1	Square tube	50 x 50 x 2.5	264

SP3	1	Circular bar	8	180
SP4	1	Flat bar	50x6	130
Pulley	1	-	-	-

The front of the base is made up of one S6 component, two S6-1s and two S6-2s, with each profile of the last two components welded perpendicular to each other at the ends of the S6. Figure 7 shows the angle of each vertex of the triangle formed (S6-1, S6-2 and the floor) and the angle of the cuts made in the profiles to align them with the floor.

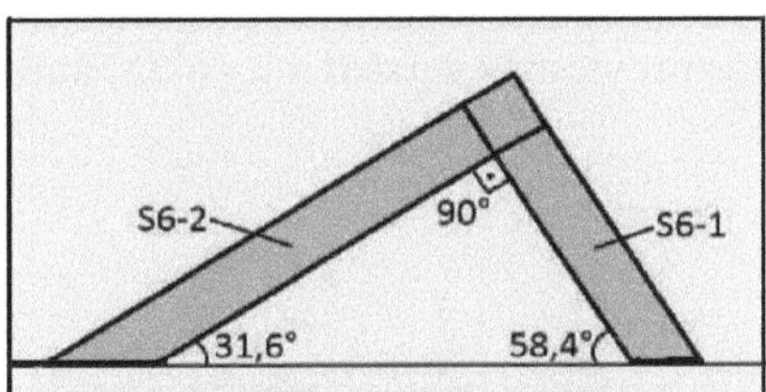

Figure 7- Angles formed by S6-1, S62
Source: CERQUEIRA; SANTOS, 2013.

The rear part is made up of items S21, S30, S31, S33, S34 and S35. The axis of rotation, which coincides with the value of the local latitude, is perpendicular to profile S21, so the angle of inclination of this profile is analogous to the local latitude, i.e. in this case the profile is inclined at an angle of 80° to the ground in a northerly direction. The other items are used to support the base and make it firmer. In order to weld S30 and S31 together, cuts were made in both parts to enable them to be attached to S21 and the assembly to be tilted latitudinally. Figure 8 shows the assembly seen from behind with the angles of each vertex.

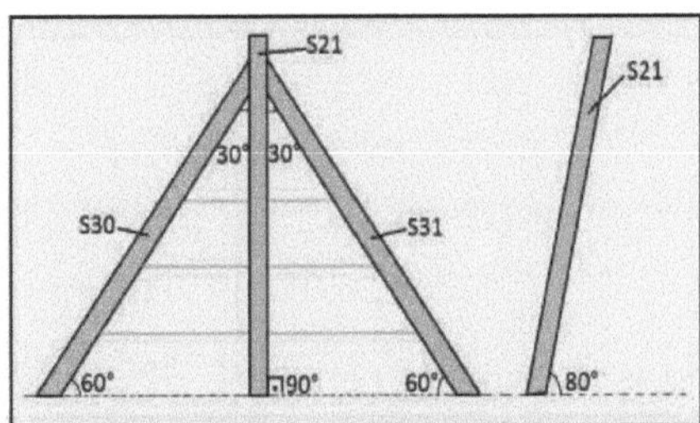

Figure 8 - Angles formed by S30, S31 and S21
Source: CERQUEIRA; SANTOS, 2013.

The ends of the base are joined by profile S2, which has been welded to the centre of S6 and the inner wall of S21. The adjacent pieces in the middle are profile S37, which is below S2 to support the base, and profile S5, which is above S2, parallel to item S21.

Above the S5 and S21 profiles, the S8 and S9 assemblies are welded together to fix the rotation support to the base. This is done using 10 mm screws belonging to the axis of rotation. Figure 9 explains the configurations previously carried out on S8, and Figure 10 shows this assembly on S5.

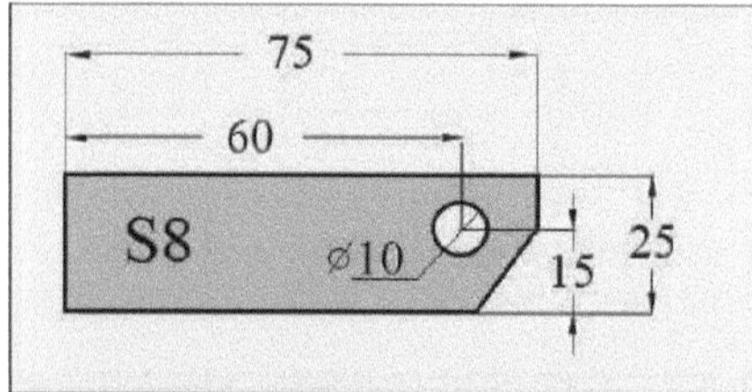

Figure 9- S8 configuration
Source: CERQUEIRA; SANTOS, 2013.

Figure 10- Set S8 and S9 on S5
Source: CERQUEIRA; SANTOS, 2013.

The assembly that forms the pulley support is made up of parts SP1, SP2, SP3 and SP4 and is located on the left wall of S2 just above profile S37. Figure 11 shows the assembly welded to S2.

Figure 11- Pulley assembly welded to S2
Source: CERQUEIRA; SANTOS, 2013.

Finally, the "hole plug" pieces were welded on, which, as the name implies, serve to seal the ends of the profiles, with the aim of preventing water from entering them and thus prolonging the useful life of the apparatus. The parts were welded together at: S10 at the front end of S2, at the tips of S6 and at the bottom ends of S21 and S37; S40 at S6-1; S41 at S6-2; and S39 at the ends of S30 and S31. However, parts S39 and S41 serve as base adjusters by means of screws. Figure 12 a) shows part S39 on S31 and Figure 12 b) shows part S41 on S6-2. Figure 13 shows the finalised base structure.

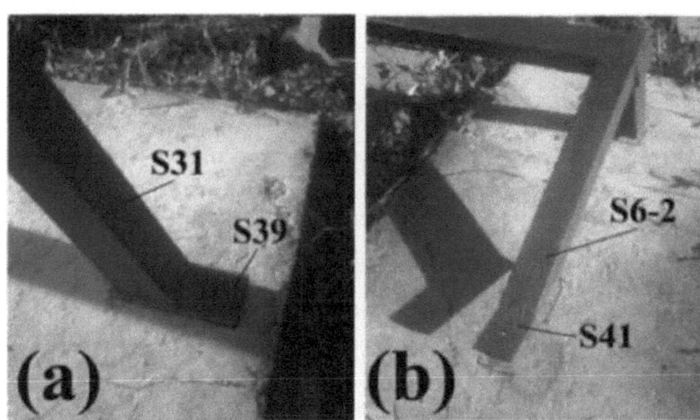

Figure 12 - Welding the parts a) S39 welded to the base; b) S41 welded to the base
Source: CERQUEIRA; SANTOS, 2013.

Figure 13 - Finished concentrator base
Source: CERQUEIRA; SANTOS, 2013.

3.1.2 Rotation support angle

The purpose of the rotation bracket template is to assist in the construction of the rotation bracket, ensuring that the parts of the bracket are precisely matched to their proper locations. Figure 14 illustrates the finished bracket template.

Most of the material used to make the template was obtained from junkyards, with the exception of the 50 by 50 mm square tubes of galvanised metallic steel, which were purchased from a specialist shop. The list of parts used to build the template is shown in Table 3, with their nomenclature and dimensions.

Figure 14- Rotation support template
Source: CERQUEIRA; SANTOS, 2013.

Table 3 - Parts list for the rotation bracket template

Nomenclature	Quantity	Profile	Dimensions (mm)	Length (mm)
GR 1	1	Square tube	50 x 50 x 2	920
GR2	1	Square tube	50 x 50 x 2	1680
GR3	2	Square tube	50 x 50 x 2	120
GR4	2	Square tube	50 x 50 x 2	100
GR 5	2	Corner	38 x 38 x 3	100
GR6	1	Corner	38 x 38 x 3	150
GR 7	1	Corner	38 x 38 x 3	264
GR 8	1	Corner	38 x 38 x 3	40
GR9	2	Flat bar	50x6	200
GR 10	1	Flat bar	50x6	50
GR 11	1	Flat bar	25x6	50
GR 12	1	Flat bar	25x6	75
GR 13	1	Corner	38 x 38 x 3	82
GR 14	2	Round bar	6	150
GR 15	2	Round bar	6	100

Before starting the welding stage of the template profiles, 8 mm holes were drilled in certain parts. Figure 15 shows these parts, which were GR5, GR6, GR10, GR11 and GR12, as well as the location of their respective holes. The accuracy of these holes is intrinsic to the accuracy of the rotation support parts and the rotation axis.

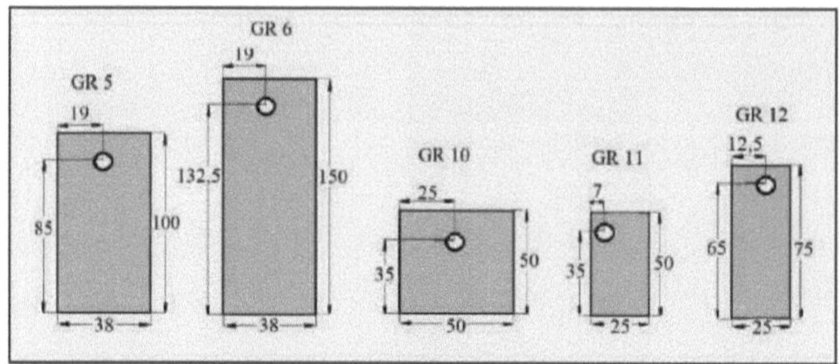

Figure 15 - Location of holes in GR5, GR6, GR10, GR11 and GR12
Source: CERQUEIRA; SANTOS, 2013.

The welding stage of the template began by adding GR1 to the centre of GR2, so that one end (the front end) of GR1 was 80 mm from the front face of GR2. Next, item GR6 was welded to the front end of GR1, according to the coordinates in Figure 16, as were the GR5 elements on the edges of GR2, taking care to keep the centre of the holes aligned with the centre of the square profile. Both of these elements served as a support for the exact insertion of the three rotation support profiles that are attached directly to the reflector.

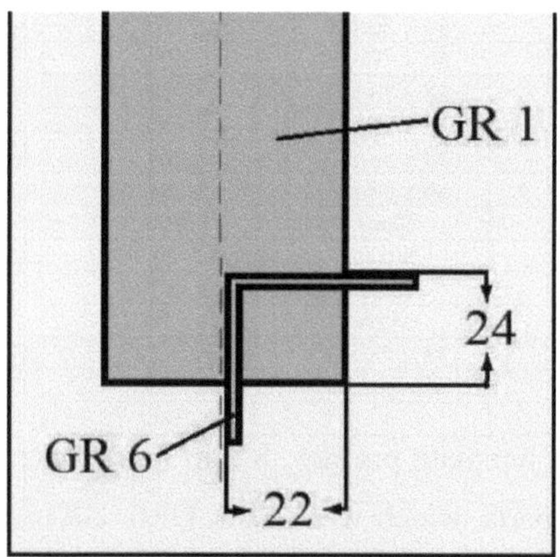

Figure 16- Location of GR6 in GR1
Source: CERQUEIRA; SANTOS, 2013.

The assemblies involving the GR3, GR4 and GR9 profiles were coupled to GR2 via GR3, so that the surfaces of both remained close. The location of the assembly is 240 mm from the edge of GR2. Here you can see the position of each piece to be welded in the assembly (Figure 17).

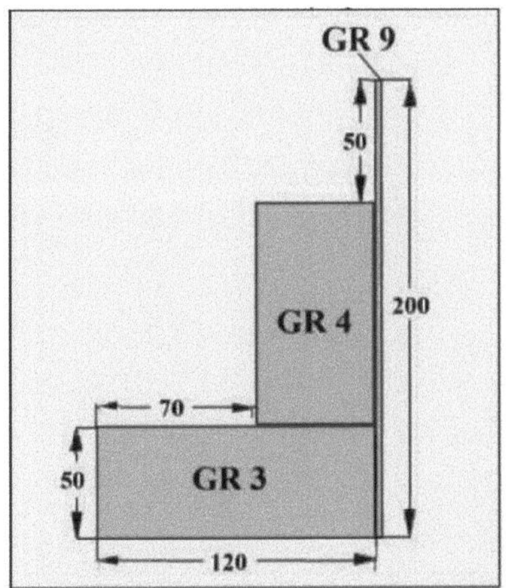

Figure 17 - Positioning of parts GR3, GR4 and GR9
Source: CERQUEIRA; SANTOS, 2013.

The GR10 and GR11 parts were welded so that the centre of their holes coincided with each other and aligned with the centre of the GR2 profile. The GR12 centre is fixed 100 mm in front of GR10 and 10 mm from the right side of GR2. The plan of the rotation support template can be visibly visualised in Appendices A, B and C.

The set of GR7, GR8 and GR13 angles is used to help install a part responsible for the counterweight system of the rotation support. Located 130 mm in front of GR10, it was inserted into the rest of the jig through GR8 in GR2, so that both surfaces remained aligned. Looking at the assembly vertically, the positioning of parts GR8 and GR13 are 90 mm and 125 mm from the bottom tip of GR7 respectively.

Finally, after all the elements of the rotation support template had been incorporated, the two 150 mm round bars (GR14) were welded to the centre of the GR2 element ends and another two 100 mm round bars (GR15) were welded to the centre of the GR1 ends. After welding, a line was tied between the two GR15s at 35 mm from the surface of GR1 and another line between the GR14s at 85 mm from the surface of GR2 (Figure 18). These perpendicularly intersecting lines were used to analyse whether the holes were precisely coincident, especially the line on GR1, which represents the axis of rotation of the entire concentrator.

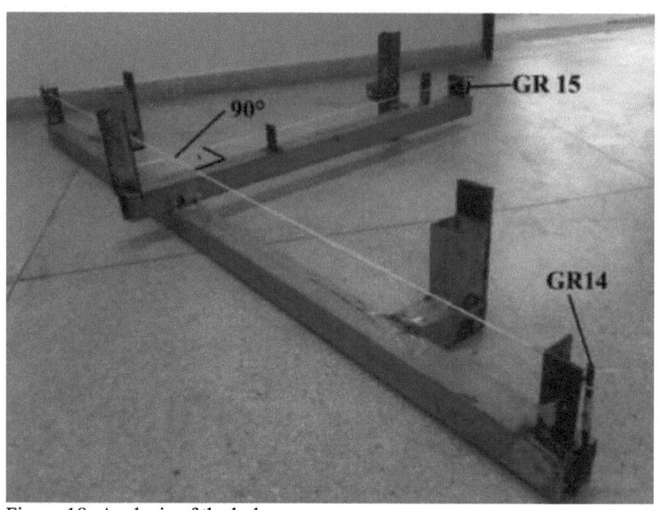

Figure 18- Analysis of the holes
Source: CERQUEIRA; SANTOS, 2013.

3.1.2 Rotation support

The rotation bracket is designed to rotate the reflector around an axis parallel to the polar axis, which coincides with the latitude angle of the location where the concentrator will be installed.

As with the rotation support template, some of the materials used were purchased from scrap yards, which means that the actual composition of some of the metals used is unknown, with the exception of the 50 by 50 mm square tubes, which are galvanised metalon steel profiles. The elements used in the support are listed in Table 4 with their respective dimensions.

Table 4 - Parts list for the rotation stand

Nomenclature	Quantity	Profile	Dimensions (mm)	Length (mm)
R1	1	Square tube	50 x 50 x 2	681
R2	1	Square tube	50 x 50 x 2	1588
R3	1	U profile	50 x 25 x 3	1360
R4	1	Corner	15x 15x3	480
R5	1	Flat bar	50x6	65
R6	2	Flat bar	50x6	180
R7	1	Flat bar	50x6	265
R8	1	Flat bar	50x6	75
R9	1	Flat bar	25x6	50
R10	1	Flat bar	25x6	25
R11	1	Flat bar	25x6	490
R12	1	Flat bar	25x6	550

R13	2	Flat bar	25x6	40
R14	1	Round bar	10	340
R15	9	Round bar	6	~15

In order to start making up the rotation support, we first had to make some configurations on certain parts, as follows: a 45° cut at one end of R1, where R5 was welded; the shaping of R3 and R4 into a semi-circumference, respectively, with radii of 400 mm and 150 mm, the centre of both coinciding with the axis of rotation of the support and 8 mm holes in components R6, R7, R8, R9, R10 and R12 as shown in Figure 19 , and 6 mm holes in the centre of R13, components that were welded to the ends of R3.

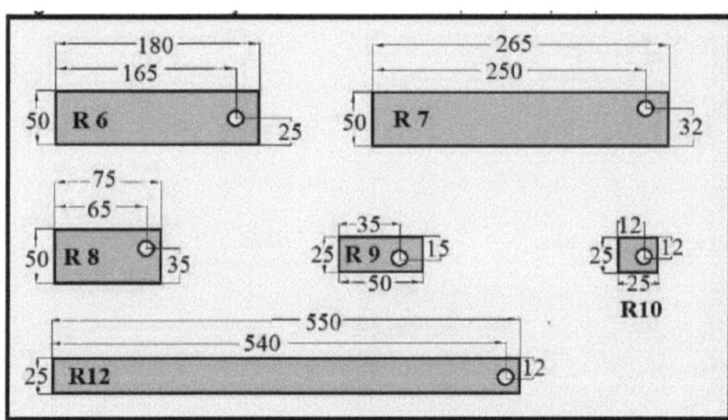

Figure 19 - Location of holes in R6, R7, R8, R9, R10 and R12
Source: CERQUEIRA; SANTOS, 2013.

After formatting the aforementioned components, the construction of the support began by fitting items R1 and R2 onto the template. The bracket and template parts were screwed together in the following order: R6 on GR5; R7 on GR6; R9 in front of GR11; R10 on GR12 and R8 on GR10. Once installed, these bracket parts were welded to R1 and R2, noting that part R7 was located exactly in the middle of R2.

Next, after welding item R3 to R2, the front assembly of the support was installed using profile R7. To do this, item R11 was first welded to profile R7 and to the front wall of R3 so that it was perpendicular to the axis of rotation. Next, part R12 and round bar R14 were welded simultaneously to profiles R7 and R11, so that the centre of the hole in R12 was close to the edge of R11.

Finally, the R4 semi-circumference was attached to the upright with the help of the GR7, GR8 and GR13 sets from the support template and nine R15 pieces, approximately 130 mm long. An important detail is that, like the R3 piece, the centre of both is coincident with the concentrator's axis of rotation. Figure 20 shows the support from the rear, while Figure 21 shows a top front view.

Figure 20 - Completed rotation bracket seen from behind
Source: CERQUEIRA; SANTOS, 2013.

Figure 21- Rotation support completed, front view
Source: CERQUEIRA; SANTOS, 2013.

3.1. 4Seasonal Adjuster

Seasonal adjusters have the function of conforming the surface and inclination of the reflector according to the seasonal variation resulting from the earth's translation movement, ensuring that the focus remains fixed throughout the year.

Each adjuster basically consists of a round tube, a round bar that slides inside the tube, and a screw lock that holds the bar to the tube. The materials used for their construction are listed in Table 5.

Table 5 - Seasonal adjuster parts list

Nomenclature	Quantity	Profile	Dimensions (mm)	Length (mm)
AS 1	1	Circular tube	18	500
AS2	1	Circular tube	18	700
AS3	1	Circular bar	10	500
AS4	1	Circular bar	10	700
AS5	2	Circular bar	6	100
Hexagonal bolt with nut	6		8	80

To assemble the adjuster locking system, the following steps were taken: first, a hole corresponding to the diameter of the bolt (8 mm) was drilled 40 mm from one end of each tube; then a nut was welded into each hole; the AS5 piece was fixed onto the head of each bolt, making up the lock. Finally, the remaining four bolts were welded to the other ends of the tubes and to one end of each bar.

The adjusters are located between the centre bar and the rotation support, with the 50 mm adjuster at the top and the 70 mm adjuster at the bottom of the concentrator. Figure 22 shows the two completed adjusters.

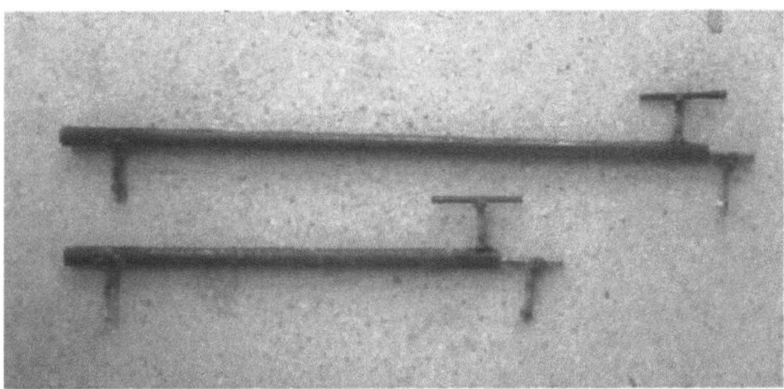

Figure 22- Seasonal adjusters
Source: CERQUEIRA; SANTOS, 2013.

Seasonal adjustment is carried out manually, simply every 3 or 4 days, or when a change in focus direction is observed. For this procedure, with the adjuster locks released, the reflector is free to move within the limits of the adjusters, and can be contracted or flattened, until the focus is visually found and the adjusters are locked again.

3.1.5 Building the Reflector Panel

The reflector panel is the most important part of the solar concentrator, as it delimits the area of the paraboloid responsible for converging solar energy towards the focus. Its structure, made up of steel profiles, is both rigid and flexible, so its surface can be conformed to each seasonal adjustment.

The reflector is attached to the concentrator assembly via parts BC3 and BC2 of the centre bar and parts BM4 of the arc. The construction of the reflector was divided into five stages: the acquisition of tools, the construction of the cross bars, the centre bar, the arch and the fixing of the mirrors.

3.1.2.1 Tools

To build the reflector, it was first necessary to acquire and produce some specific tools, such as a piece of plywood, which served as a template for projecting the pieces, a compass for drawing the pieces with their respective radii and two steel spanners for forming the bars.

It is recommended that 1800 mm X 2450 mm X 15 mm plywood be used to make the template, as it is possible to project the reflector as a whole. However, due to the lack of material with these dimensions, the template for the parts was improvised on 1200 mm X 2200 X 15 mm plywood, thus projecting only half of the reflector's arc. However, as the two sides of the prototype are symmetrical, it was not so difficult to build the reflector.

To draw the pieces on the template with their appropriate radii (Figure 23), a compass was improvised from a 3000 mm long aluminium curtain frame. A dry point was attached to one end of the compass to hold it in the centre of the radii, and the other end was a mobile holder for a pencil or other instrument to mark out the different radii. To illustrate the BM2 parts of the reflector's edge, a string was carefully used as a compass, as it has a radius of 2970 mm for one half and 3700 mm for its other half.

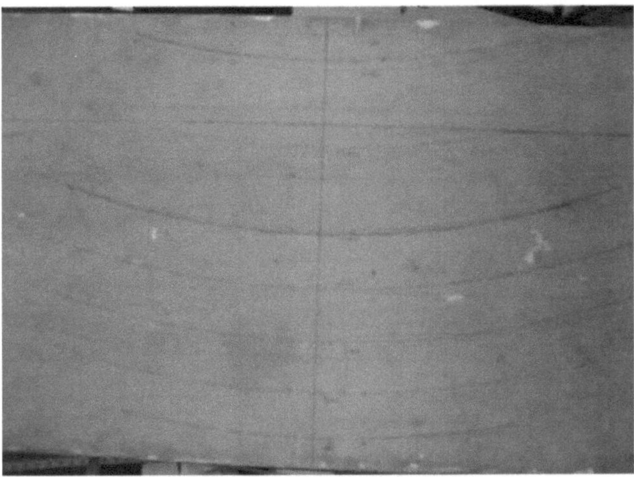
Figure 23 - Wooden plywood used to draw the template
Source: CERQUEIRA; SANTOS, 2013.

For moulding the bars, a pair of steel spanners with a 15 mm opening and a 600 mm long handle was constructed (Figure 24).

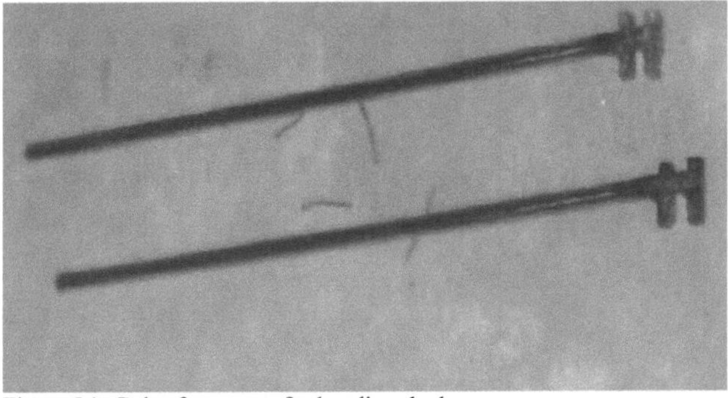
Figure 24 - Pair of spanners for bending the bars
Source: CERQUEIRA; SANTOS, 2013.

Only after acquiring these instruments was it possible to proceed with the construction of the reflector components.

3.1.5.2 Centre bar

Similarly, the centre bar is the backbone of the reflector, supporting the set of cross bars and the reflector arch. This bar is used for seasonal adjustment, i.e. the deformation of the reflector surface with each change of season. Through the central bar, the reflector is fixed to the assembly at three points: two points through the seasonal adjusters and one point in the centre.

The list of parts that make up the centre bar assembly is shown in Table 6. The template for the centre bar was drawn on the plywood according to the measurements given (Figure 25). Points 1 to 7 on the template correspond to the locations where the crossbars will be located and points 0 and 8 correspond to the arch. The lines drawn between the points coincide with the internal surface of part BC1.

Table 6 - List of centre bar parts

Nomenclature	Quantity	Profile	Dimensions (mm)	Length (mm)
BC1	1	Square bar	10	2340
BC2	1	Square bar	10	50
BC3	2	Flat bar	4 x 38	65

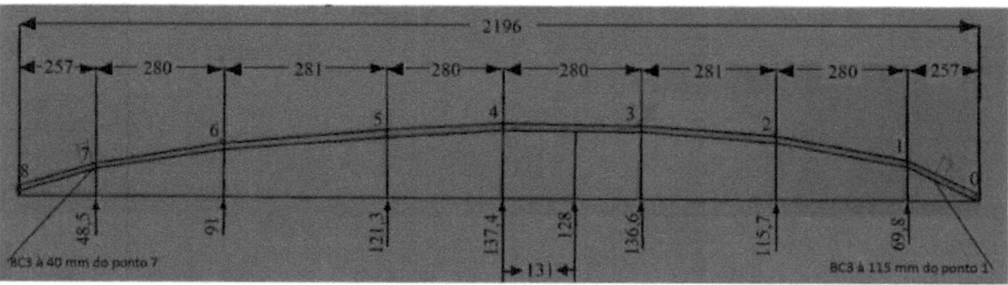

Figure 25- Measurements of the centre bar
Source: CERQUEIRA; SANTOS, 2013.

The modelling stage of the centre bar started from point 4, which coincides with the middle of BC1. The centre bar was then bent using the keys until it reached the other points. There was no need to bend or cut the scraps after points 0 and 8, as it served as a support for the arch. Figure 26 shows the centre bar aligned with the template.

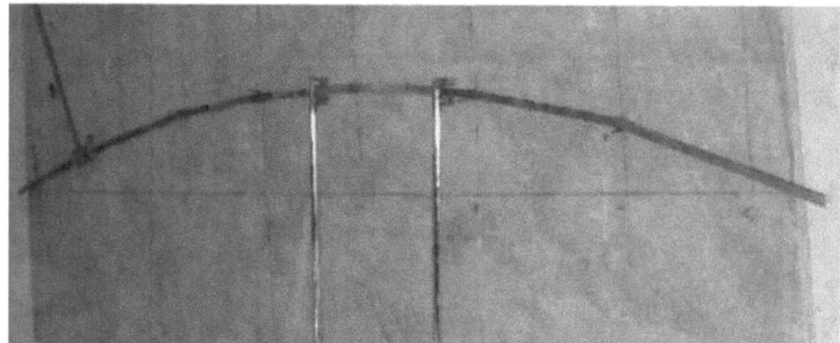

Figure 26- Moulding of the centre bar
Source: CERQUEIRA; SANTOS, 2013.

At 131 mm from point 3, towards point 4, there is a point labelled S, located on the inside of BC1. This marking coincides with the halfway point of BC2 and represents the place where it will be welded to BC1 to form the rotating support fitting. Before welding, each piece was first cut to a length of 20 mm and a depth of 5 mm with a sander in order to weld them together and obtain the socket (Figure 27). After the cutting and welding stage, BC1 was checked and corrected in the template due to non-conformities with all the points, caused by the thermal expansion of the process.

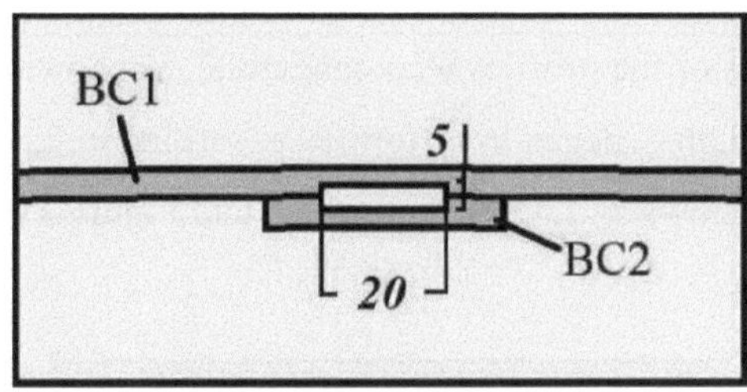

Figure 27- Part CB2 welded to the centre bar
Source: CERQUEIRA; SANTOS, 2013.

The two BC3 parts were drilled 8mm holes as shown in Figure 28 and then welded to BC1. To weld the BC3 parts, BC1 had to be clamped in a vice. The centre of one of the BC3 parts was tangential to a point located between points 0 and 1, exactly 115mm from point 1, and the centre of the other BC3 was tangential to a point located between points 7 and 8, specifically 40mm from point 7. Again, after welding, corrections had to be made to BC1 due to the thermal expansion caused by the process.

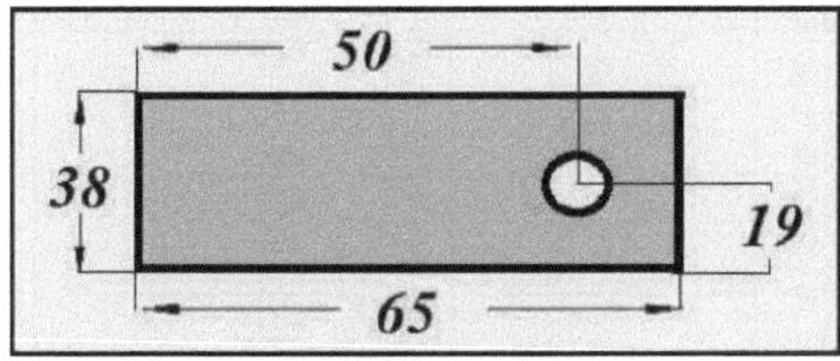

Figure 28- Location of BC3 borehole
Source: CERQUEIRA; SANTOS, 2013.

3.1.5.3 Crossbars

The projections of the transverse bars were made on the opposite side of the arch template and the centre bar. Their function is to support the

aluminium that support the mirrors, so that they remain coincident with the surface of the paraboloid.

Table 7 shows the materials used to build the crossbars, as well as their nomenclature and the radii of each piece. Each crossbar was made up of three round bars, two 6mm and one 8mm.

The larger diameter profiles are the centrepieces of the crossbar, and the other two smaller diameter profiles are welded at their ends. Welding was carried out with the profiles still straight and so that the ends of the profiles were tangential, as shown in Figure 29. Once the bars had been welded, they began to be formed according to the scratches made on the template. Figure 30 shows an LM being formed, and Figure 31 shows all the LMs moulded on the template.

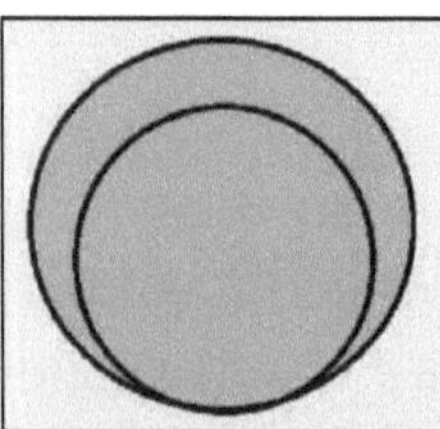

Figure 29-Tangency between two pieces of the same BT
Source: CERQUEIRA; SANTOS, 2013.

Table 7 - Cross member parts list

Nomenclature	Quantity	Profile	Dimensions (mm)	Length (mm)
BT 1	2	Circular	6	256
BT 1	1	Circular	8	514
BT2	2	Circular	6	325
BT2	1	Circular	8	639
BT 3	2	Circular	6	392
BT 3	1	Circular	8	785
BT4	2	Circular	6	406
BT4	1	Circular	8	811
BT 5	2	Circular	6	390
BT 5	1	Circular	8	781
BT6	2	Circular	6	346
BT6	1	Circular	8	692
BT 7	2	Circular	6	251
BT 7	1	Circular	8	500

Figure 30- Crossbar being formed
Source: CERQUEIRA; SANTOS, 2013.

Figure 31 - Crossbars already moulded on the template
Source: CERQUEIRA; SANTOS, 2013.

3.1.5.4   Reflector arc

The arc elliptically represents the limits of the parabolic segment, so its front inner edges tangent with the surface of the paraboloid. Through the arch, the reflector is attached to the concentrator assembly by the two BM4 parts. Table 8 shows the quantity, profiles and measurements of the parts used to build the reflector arch.

Table 8- Parts list for the reflector arch

Nomenclature	Quantity	Profile	Dimensions (mm)	Length (mm)
BM1	2	Square bar	10	1840
BM2	2	Square bar	10	650
BM3	1	Square bar	10	1200
BM4	2	Flat bar	4x38	150
BMe	14	Flat bar	12 x 3	30

Due to the lack of 1800 mm X 2450 mm plywood that would fit the design of the reflector as a whole, only half of the arch was drawn on 1200 mm X 2200 mm plywood. There weren't many difficulties when assembling the parts, as the two halves are similar.

The template was drawn using lines, strictly following the measurements and radii, as shown in Figure 32, with the exception of the BM2 parts. Unlike the other parts, which have a curvature towards the inside of the reflector, the BM2 parts have a curvature towards the back of the reflector. The BM2s have two different radii (Figure 33): half of the part towards point 1 has a radius of 2970m and the other half, towards point 3, has a radius of 3700mm.

As the template lines represent the limits of the paraboloid and therefore belong to its surface, the parts were shaped so that the internal edges in contact with the template coincided with the lines. In order to mould the parts perfectly, in addition to using the bending keys, it was also necessary to hammer the profiles a few times, especially at their ends.

Once all the profiles had been moulded, the welding stage began. Firstly, BM1 and BM2 were welded together, but as there was only the right side of the reflector in the template, BM1 was welded to BM2 normally, while BM1 and BM2, which make up the left side of the reflector, had to be welded "upside down", i.e. with their edges in contact with the wood facing upwards. Once each welding stage had been completed, the profiles were consulted and some corrections made to keep them in line with the template.

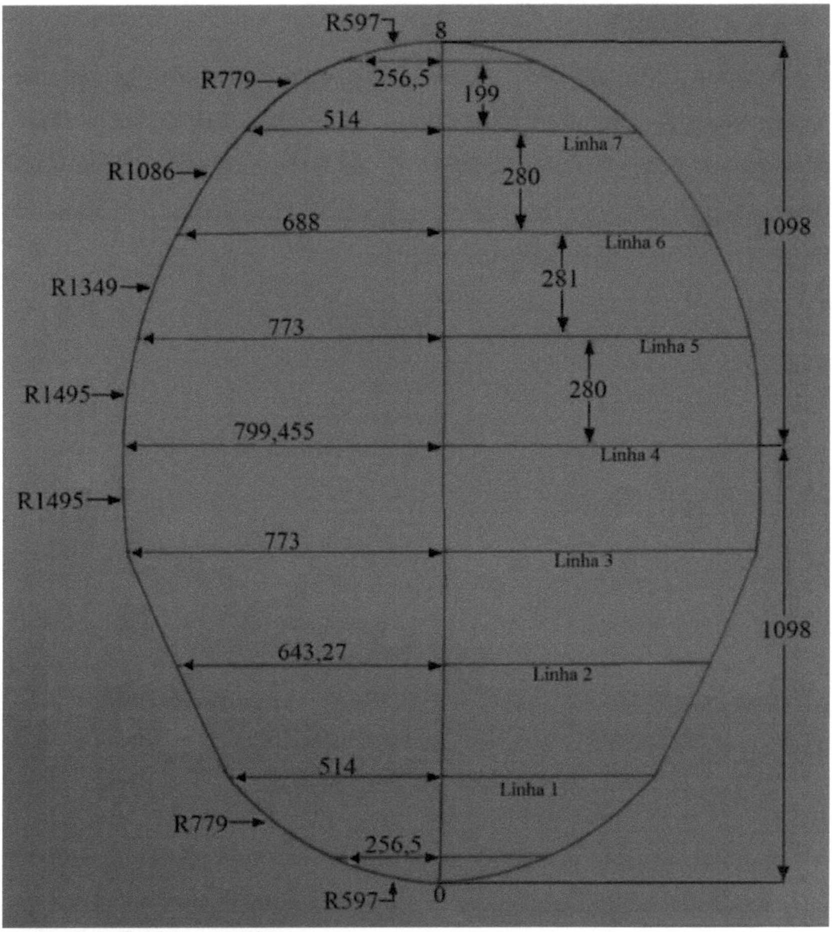

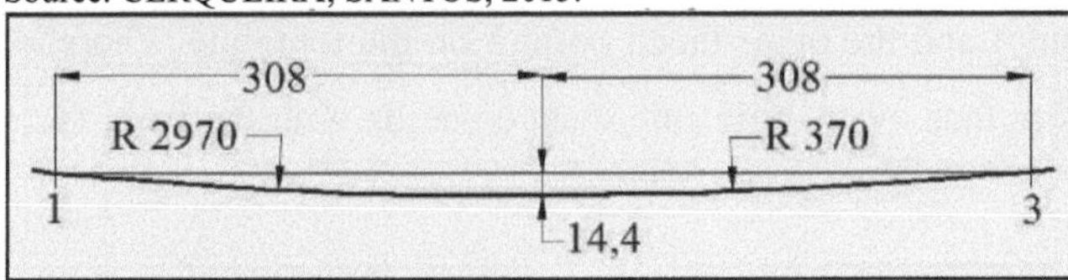

Figure 32- Reflector arc template
Source: CERQUEIRA; SANTOS, 2013.

Figure 33- Template for part BM2
Source: CERQUEIRA; SANTOS, 2013.

The BM1s were then welded together. Firstly, the other half of the reflector was drawn on another piece of wood to help keep the arc coincident with the surface of the paraboloid and thus guarantee maximum reflectance of the reflector on the target. Once the parts had been aligned and welded, corrections were again made to the parts due to non-conformities caused by thermal expansion from the high welding temperatures.

Before inserting BM3, the BM1 and BM2 sets were fastened to the template with nails to prevent them from misaligning during the welding of BM3. To fix BM3, some wooden supports were used underneath BM2 and BM3, so that the edges of the ends of both were tangential. Only after this process was BM3 welded to the assembly. Figure 34 illustrates the arc welded onto the template.

Figure 34- Welded arch

Source: CERQUEIRA; SANTOS, 2013.

Before inserting the BM4 parts, they were drilled with an 8 mm hole, with the centre of the hole located 9 mm from one of the sides and 17 mm from the top of the bar, as shown in Figure 35. The BM4 bars were welded so that the lower edge of the hole side coincided with point 4 and the other faced point 5 on the template. There is also the detail that the opposite face of the hole side must coincide with the front face of the reflector arch.

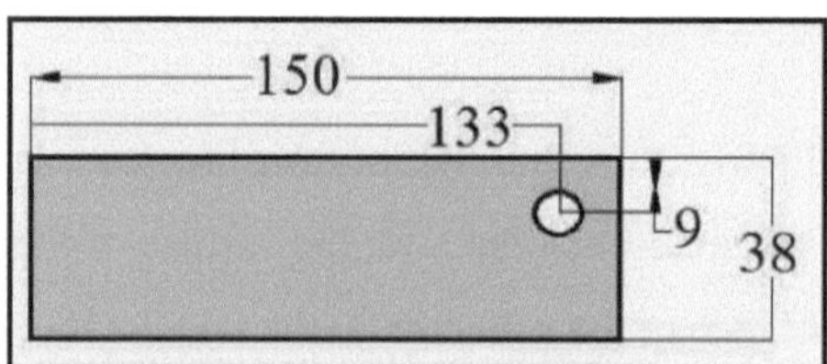

Figure 35- Location of the hole in part BM4
Source: CERQUEIRA; SANTOS, 2013.

Lastly, the BMe pieces were welded together to join the crossbars to the arch and keep them coincident with the surface of the paraboloid. The BMe pieces are 30 mm long, 10 mm of which were welded in contact with the front face of the reflector so that they were aligned with the lines referring to the cross bars (Figure 36).

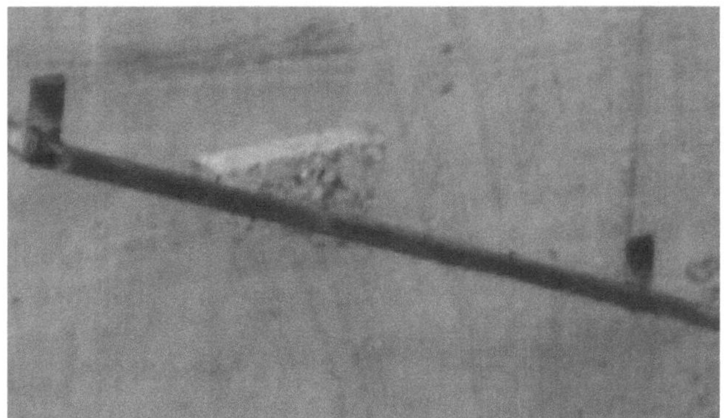

Figure 36 - BMe parts welded in the reflector arc
Source: CERQUEIRA; SANTOS, 2013.

3.1.5.5 Reflector assembly

Once all the bars had been built, the reflector assembly stage began. At first, markings were made on the centre bar for the points where the cross bars and the arc would be welded. The reflector arc was welded onto the centre bar (at points 0 and 8) so that the arc was in front of the centre bar (Figure 37). Before final welding, it was ensured that the centre bar was exactly aligned with the middle of the solar panel.

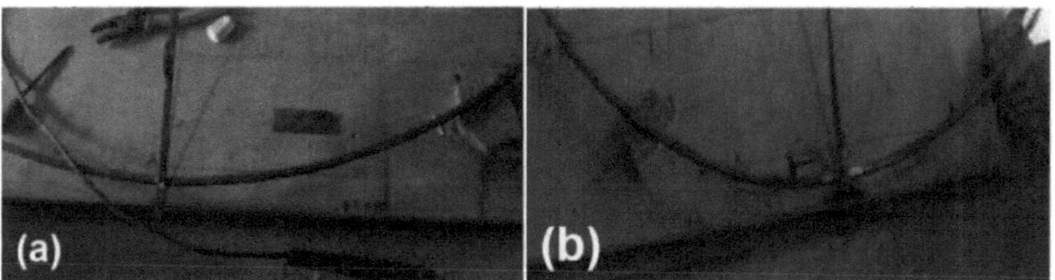

Figure 37 - Insertion point
(a) Centre bar on the arc, point 0; (b) Centre bar on the arc, point 8
Source: CERQUEIRA; SANTOS, 2013.

The process of welding the transverse bars in the "backbone" began with the BT4 (centre) piece, which was precisely aligned halfway along the centre bar and then carefully welded (Figure 38). The next step was to weld its ends to the BMe parts of the arch. To do this, the centre bar had to be pressed and the height of its centre in relation to the front face of the arc checked, which corresponds to 137.7mm.

Figure 38 - BT4 and BT5 welded to the arch and centre bar
Source: CERQUEIRA; SANTOS, 2013.

To assemble the other crossbars in the reflector, the same steps were followed as when BT4 was inserted. BT3, BT5, BT6, BT7, BT2 and BT1 were inserted into the reflector assembly respectively. However, each BT has a different relative height from the vertical centre of the central bar to the front face of the reflector arch, as described in Table 9. Figure 39 shows the completed reflector structure.

Table 9 - Distance between the centre of the transverse bars and the front face of the arch

Crossbar	Height between the front face of the arch and the centre of the Cross Bars (mm)
BT1	68
BT2	113
BT3	136
BT4	137,7
BT5	120
BT6	91
BT7	45

Figure 39- Assembled reflector skeleton
Source: CERQUEIRA; SANTOS, 2013.

Once the procedures described had been carried out, the reflector structure was finalised, all that remained was to paint it and check that the distances between the crossbars and the front face of the arch were in accordance with Table 9 (Distance between the crossbars and the front face of the arch), when it was mounted on the rotation support. It was necessary to adjust by pulling and/or pushing the centre bar until they all reached the desired value.

3.1.5.6 Fixing the mirrors

The installation of the mirrors was characterised as the last stage in the construction of the reflector panel. The mirrors take up exactly the surface of the segment of the parabola in revolution, i.e. it is on them that the sunlight will be directed towards the focus.

In this process, a total of 16 aluminium profiles 3mm thick and 25mm wide were used, but of varying lengths depending on their position on the reflector. Around 350 ordinary mirror plates with a reflectance of 70 per cent were used. Basically all the mirrors had dimensions of 2mm X 75mm X 100mm in terms of thickness, width and length, except for a few pieces that were cut to size for the panel's edges. All the edges were painted black to reduce light reflection from the surroundings. Figure 40 shows the materials used in this phase.

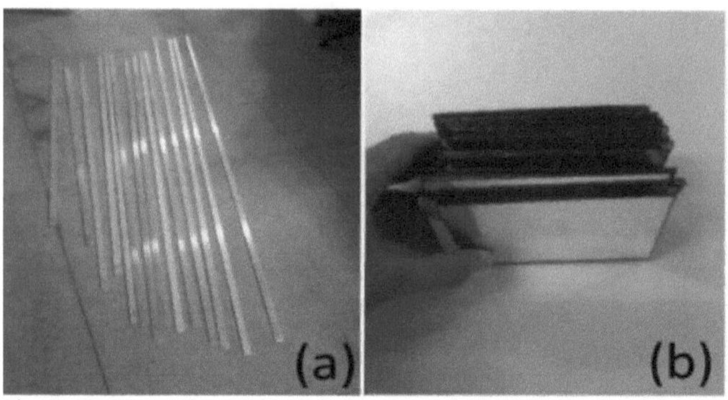

Figure 40 - Materials used in the reflector panel (a) Aluminium bars; (b) Mirrors (75 mm x 100 mm) with painted sides

Source: CERQUEIRA; SANTOS, 2013.

The mirrors were mounted from the inside to the outside of the panel, so that the first aluminium bar was as close as possible to the central bar. To fix the profiles, they were placed on the panel so that they extended 200 mm over the reflector arch. Markings were then made close to the points where the aluminium bar rested on the transversals and the arch.

These markings were centred on the width of the profiles and drilled with a 3/16" drill bit. The profiles were then provisionally attached to the reflector with 2 mm diameter coated copper wire to outline the holes that would hold the mirrors.

With the aid of a mirror that had already been dimensioned, the position of the next bars, which should be arranged parallel to each other, was traced out, with the distance between the edges of a mirror corresponding to the outer edges of the bars (Figure 41). Only after the first two bars had been installed was it possible to insert the first column of mirrors.

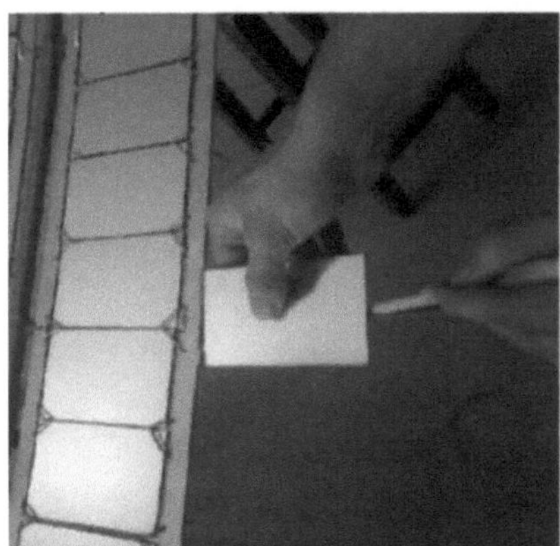

Figura 41   - Marking for fixing the next aluminium bar
Source: CERQUEIRA; SANTOS, 2013.

The rectangular mirrors were placed horizontally on the reflector so that their vertices were as close to the centre of the aluminium profiles as possible. The first mirror was positioned over BC4, so that from a frontal view it appears that the crossbar passes through the middle of the mirror, as shown in Figure 41. The other mirrors were placed side by side with a spacing of 1.5 mm to allow the wires that hold them to pass through. Markings were then made on the bar, approximately 12 mm from the vertices of the mirrors.

These markings, also centred on the width of the aluminium profiles, were drilled with 3 mm drills to pass through the wires that hold the mirrors. Due to the greater curvature of the lower part of the reflector, six holes were also drilled between BC3 and the arch with an 8 mm drill bit, to make the reflector more flexible when making seasonal adjustments.

Once all the holes had been drilled, the openings were sanded and the aluminium bars were permanently attached to the reflector with 2 mm coated rigid copper wire, purchased from a building shop. Next, the mirrors were fixed with the same type of wire, only 1 mm in diameter, found inside mains cables. Both sides of the centre bar evolved gradually in a similar way, i.e. as a row of mirrors was placed on one side, the same was done on the other. Figure 42 shows two columns of fixed mirrors.

Figura 42   - Mirror columns fixed to the reflector
Source: CERQUEIRA; SANTOS, 2013.

This procedure was carried out until the mirrors were installed. However, because of the curve of the reflector, as soon as an aluminium bar was inserted, it had to be adjusted so that it was flush with the dish. To do this, it was necessary to build a tool that would allow them to bend sideways without deforming their facets. This tool is based on four pieces of 25mm X 25mm X 50mm angle brackets, screwed onto a board above an iron "bar", with spacing between them just to run the profile and another for hammering (Figure 43). The aluminium bars were then adapted by hand until they reached their respective places.

Figure 43- Aluminium bar on the instrument for adaptation
Source: CERQUEIRA; SANTOS, 2013.

3.1.6 Assembly and adjustments

Once the structural parts that make up the solar concentrator (base, rotation support and reflector) had been built and painted, a preliminary assembly was carried out to check whether any corrections needed to be made.

For these corrections, a levelled pavement had to be built at the location where the concentrator was to be installed, due to the unevenness of the ground, in order to keep it as firm and fixed to the ground as possible, guaranteeing the precise permanence of the axis of rotation and therefore of the focus.

Just to emphasise, the axis of rotation of the reflector is parallel to the polar axis and coincides with the latitude angle of the location. It was therefore essential to pay attention to two details: the positioning of the prototype, which should be directed in front of the Equator; and the precision of the inclination of the axis of rotation. With the aid of a GPS, the concentrator was properly orientated to geographic north (Figure 44).

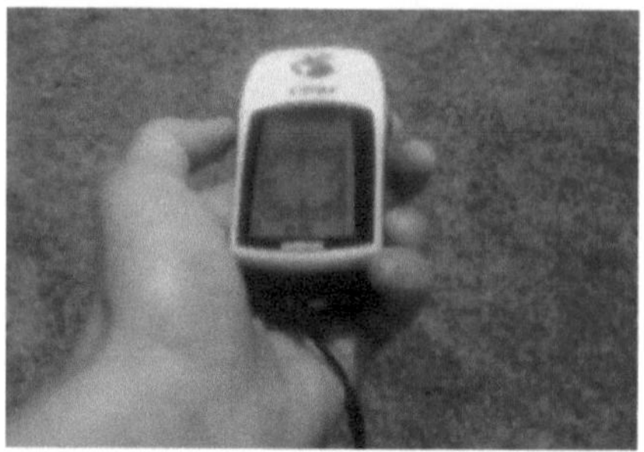
Figure 44 - Compass <u>directing</u> north
Source: CERQUEIRA; SANTOS, 2013.

To check the accuracy of the axis of rotation, a straight bar was passed between the holes in the S8's, then a plumb bob was attached to the centre of the bar, and finally the angle between the straight lines (of the bar and the plumb bob) was checked with a protractor ruler, which should be between 80° and 81°. Figure 45 shows this procedure.

Figure 45 - Method used to check the angle of the axis of rotation
Source: CERQUEIRA; SANTOS, 2013.

Once the above assumptions had been checked and ensured, the concentrator continued to be assembled, starting with the junction of the rotation support to the base, to which the R8 and R9 parts of the support were attached with hexagonal screws with partial threads and 10mm diameter nuts to the S8 parts of the base (Figure 46).

Figure 46 - Rotation bracket on the concentrator base
Source: CERQUEIRA; SANTOS, 2013.

The reflector was then installed on the rotation support using five points, two indirect and three direct. The indirect ones through the seasonal adjusters, where the smallest (50mm) was at the top between parts R10 and BC3 and the largest, at the bottom (70mm), between R12 and BC3; the direct ones through R6's and BM4's (ends) and through R7 with BC2 (centre). Both points were connected using hexagonal screws with a total thread of 8mm.

With this part of the concentrator completed, all that remained was to assemble and attach the rotation system to make the reflector rotate around its axis automatically. Figure 47 shows the concentrator properly assembled, with rear, side and front views.

Figure 47 - Total view of the concentrator (a) Rear view; (b) Side view; (c) Front view
Source: CERQUEIRA; SANTOS, 2013.

3.2 Solar cooker

One of the aims of the concentrator developed is to assess its applicability in cooking food in a simple, convenient and efficient way. A simple solar cooker with the approximate dimensions of a conventional cooker was therefore designed.

A large part of the materials that make up the cooker were acquired in junkyards, so their exact composition is unknown. Its structure supports a grill that keeps the pot positioned at the focal point of the reflector panel. Table 10 lists the parts used and Figure 48 shows their locations.

Table 10- Stove parts list

Nomenclature	Quantity	Profile	Dimensions (mm)	Length (mm)
F1	2	Corner	25 x 25 x 2	500
F2	2	Corner	25 x 25 x 2	650
F3	2	Corner	25 x 25 x 2	640
F4	2	Corner	25 x 25 x 2	490
F5	4	Corner	19x19x2	900
F6	2	Round bar	10	490
F7	1	Round bar	10	640
F8	2	Flat bar	12x2	490
F9	4	Round bar	5	870
F10	4	Round bar	5	490

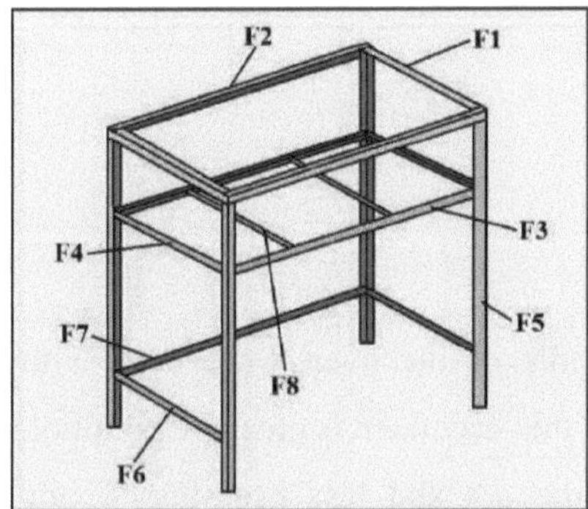

Figure 48 - Structure of the designed cooker
Source: CERQUEIRA; SANTOS, 2013.

Construction of the cooker began at the top, where the F1's were welded to the inside edges of the side angles (F2's). Afterwards, the feet that support the cooker (F5's) were inserted, fixed to the inside edges of the F1's.

Before welding the parts that support the lower boundary wall to the focus, a concrete block was used to visually analyse the height of the focus in relation to the ground surface, which corresponded to 760 mm. The F3 pieces were then welded flush against the F4 pieces, 740 mm from the lower limit of the F5 pieces. The F8 pieces were welded between the F3 pieces, supporting one of the refractory blocks. The welding process ended with the insertion of the F6's and F7's at the bottom of the cooker legs. These parts only serve to make the structure firmer. Figure 49 shows the structure of the cooker from the side, with the exception of one of the F4 parts, the F6's, F7 and F8's.

Figure 49- Stove structure seen from the side
Source: CERQUEIRA; SANTOS, 2013.

After the structure was completed, the walls of the oven were built with refractory blocks, serving as a thermal insulator. In this way there is greater absorption and concentration of the energy reflected into the pot and less heat loss to the environment.

To build the refractory plates, moulds were made with specific sizes for each wall. All the slabs were built to the same thickness (25 mm), so that they would fit into the corners of the cooker structure.

A total of four walls were built: the base (lower limit of the focus), the bottom and the two sides. The moulds were mounted on a smooth surface between rectangular pieces of wood 25 mm thick (Figure 50).

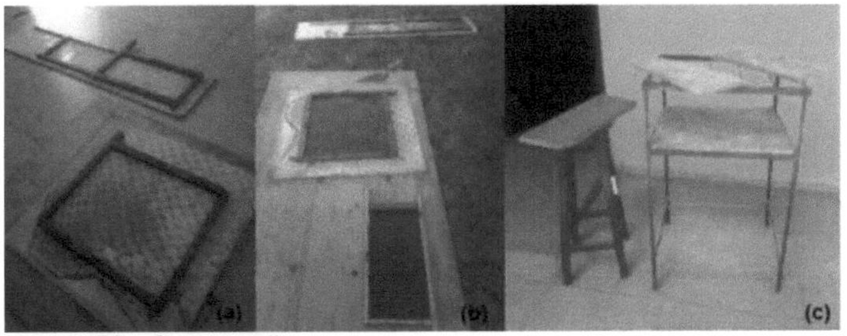

Figure 50 - Stove walls
(a) Wall moulds assembled; (b) Walls drying in the moulds; (c) Structure ready
Source: CERQUEIRA; SANTOS, 2013.

The formwork for the walls was filled in three layers: the first, 7 mm thick, was made up only of sand (65%) and cement (35%); the second, 12 mm thick, was made with a mixture of washed sand (20%), brick dust (15%), expanded clay (20%), sand (15%), refractory mortar (10%) and cement (20%) and a small amount of water; and the last layer, 6 mm thick, was filled purely with refractory mortar.After fitting the walls, the inside of the cooker was plastered, both to plug the gaps between the walls and to improve thermal insulation. At first, the intention was to use only refractory mortar, but since this was not available and it was difficult to find this material, the decision was made to use a mixture made up of 40% expanded clay, 60% clay soil, a small amount of water and 1kg of sugar. This procedure was recommended by artisans from Deodoro who make clay ovens. Figure 51 shows how the plaster inside the cooker was made.

Figure 51 - Plastering the inside of the walls
Source: CERQUEIRA; SANTOS, 2013.

Once the thermal insulation of the cooker was complete, a thin stainless steel plate was attached, following the base and bottom wall of the cooker (between the side walls) as shown in Figure 52. The purpose of this plate is to reflect the scattered light onto the cooker as well, in order to improve the concentration of heat in the cooker.

Figure 52 - Stainless steel plate attached to the inside of the cooker
Source: CERQUEIRA; SANTOS, 2013.

The final stage was the construction of the grid that supports and holds the pan directly at the focal point. Figure 53 describes its structure and dimensions. It was built in two stages, the first of which was to bend the F9's and the second was to weld them to the F10's. Figure 54 shows the finished cooker.

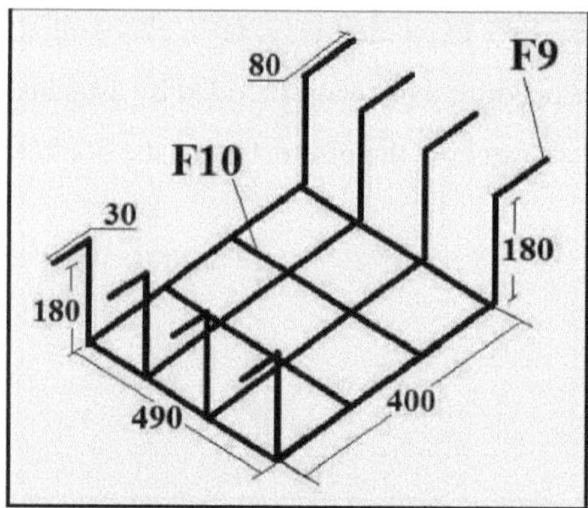

Figure 53-Drawing of the panel support structure
Source: CERQUEIRA; SANTOS, 2013.

Figure 54 - Completed cooker
Source: CERQUEIRA; SANTOS, 2013.

CHAPTER 4

APPLICATION OF THE FIXED-FOCUS PARABOLIC TROUGH SOLAR CONCENTRATOR

This chapter describes the methodology used to assess the practical usability of the solar concentrator for cooking food.

4.1 Food Cooking

Cooking is the physical-chemical process in which food is subjected to the action of heat in order to change its texture, appearance or flavour, so that it can be more easily digested and incorporated into other foods. The heat source normally used for cooking is the fire generated by burning wood or cooking gas. Cooking can be defined as raising the temperature of food sufficiently to cause irreversible changes (COENDERS, 1996).

In the solar oven, cooking was done by converging solar energy directly onto the wall of the pan, which was positioned at the focal point of the reflector panel, eliminating the need for a secondary reflector to reflect the light to the bottom of the pan. At this stage, ordinary black aluminium pans were used in order to achieve greater absorption of electromagnetic radiation.

Just to emphasise, once the pan is situated in the focus, there is no need to move it, as the reflector panel rotates according to the apparent movement of the sun, keeping the focal point constant. However, although the concentrator was designed to rotate automatically, during the tests it was rotated manually every time the focus was seen to move, so the solar tracker system still has to be installed to automate it.

The cooking tests were carried out from the 8th to the 12th of April 2013, both starting at 11am. The aim was to prepare a meal for lunch, so the tests were carried out over a five-day period so that there was a varied menu. Table 11 lists the meals prepared each day, the approximate time taken to prepare them and the average temperature each day.

Table 11 - Meals prepared in the cooking tests

Date	Meals	Preparation time (approx.)	Local temperature (°C)
08/04/2013	Spaghetti in tomato sauce with sausage	00h25min	30.5
09/04/2013	Chicken stroganoff with rice	01h25min	30.0
10/04/2013	Spaghetti in minced meat sauce with ricotta cheese	00h40min	30.0
11/04/2013	Boiled beef with rice	01.30am	29.6
12/04/2013	Beans, rice and roast beef	02h00min	30.2

Basically all the days were sunny with little cloudiness, which was favourable for using the solar cooker. With the exception of Monday (08/04) and Friday (12/04), when some denser clouds appeared and obstructed the passage of sunlight over the concentrator, slowing down cooking. One factor that interfered with the results on the 11th and 12th was the fact that the pressure cooker used had problems with the rubber sealing on the lid, allowing the pressure to escape, until a solution was improvised to seal this leak, which made cooking the beef and beans take longer.

Although each food item was cooked in turn, because the prototype cooker has only one "mouth", i.e. one focal point, it was possible to prepare different meals in a relatively short space of time, indicating that the thermal application of solar energy for cooking using this prototype is effective. Figure 55 shows the food preparation area.

Although the concentrator only converges direct solar radiation, it is possible to apply solar energy at other times by using ovens with heat storage systems. Scheffler's solar oven proved to be efficient for cooking food in both non-aqueous (dry heat) and aqueous (moist heat) environments, thus avoiding the burning of around 10kg/household/day of firewood.

Figure 55 - Cooking food
(a) Preparing stroganoff; (b) Pressure cooker at the focal point during the preparation of beans; (c) Preparing pasta

Source: CERQUEIRA; SANTOS, 2013.

CHAPTER 5

RESULTS AND DISCUSSIONS

This chapter discusses the concentration factor of this prototype solar concentrator, and analyses the temperature reached in a concrete block positioned at the focal point of the reflector, the ambient temperature of the test site (IFAL-MD) and hourly sunlight incidence values acquired by the National Meteorological Institute (INMET) for the city of Maceió/AL, the nearest city with data of this nature.

5.1 Concentration Factor and Theoretical Power

The reflector panel is the main component of the prototype and has a total area of 2.7m^2 . Because of possible inaccuracies in the reflective surface, it is appropriate to subtract 10 per cent from its total area. The concentration factor of a solar concentrator is given by dividing the reflective area by the area of the focus ($C = A_r/ A_f$).

Due to the seasonal adjustment, the useful reflective area of this model depends on its aperture factor, which is the percentage of the area that will receive the sun's rays perpendicularly and converge them towards the focus. To do this, assuming that the change in the angle of solar incidence is constant throughout the year, 23.50° is added for the summer solstice and 23.50° is subtracted for the winter solstice in the equation: (Aperture factor = cos (43.23° ± Solar inclination/2)). The value of 43.23° refers to the inclination of the dish's opening plane during the equinox (DIB, 2009). The area of the focus is determined by the equation that calculates the area of a circle ($A_f = \pi . r^2$).

Taking these assumptions into account and applying them to 15 and 16 April 2013, the aperture factor was 0.77 and the useful reflective area of this prototype was 1.87m^2 ($A_r = (2.7 - 10\%) *$ aperture factor). The focus radius was approximately 08 centimetres, so the focus area was 0.02m2. Given these values, the concentration factor of this prototype for 15 and 16 April 2013 was equivalent to 93.6.

The concentrator's reflective surface was covered with small 75mm X 100mm flat mirrors. Assuming that the reflectance of these mirrors is 80 per cent and considering the values dimensioned above, at times when the solar irradiance is 1,000 W/m2, the theoretical power received at the focal point would be 1,496 W. Estimating an average power of 900 W/m2 per hour on a clear day, the energy concentrated at the focal point between 10 am and 2 pm would be 24,235 KJ. Considering the data in Table 12, this value is equivalent to 1 m^3 of biogas.

Table 12 - Energy equivalence with biogas for various sources

1 m^3 biogas	5,000 kcal - 7,000 kcal
1 m^3 biogas	0.57 litres of paraffin
1 m^3 biogas	0.55 litres of diesel oil
1 m^3 biogas	0.45 kg of liquefied gas
1 m^3 biogas	0.79 litres of alcohol fuel
1 m^3 biogas	1,538kg of firewood
1 m^3 biogas	1,428 kwh of electricity
1 m^3 biogas	20,900 - 29,260 kJ

Source: (DEGANUTTI 2002)

5.2 Focal Point Temperature Analysis

On 15 and 16 April 2013, experiments were carried out to monitor the heating curve in a concrete block positioned at the focus of the solar concentrator. For this purpose, data was collected on the ambient temperature and the temperature at the focus every 30 minutes between the hours of 10am and 2.30pm. Between 10 a.m. and 2 p.m. the data was collected with the spotlight on the concrete block, while at 2.30 p.m. the temperature of the block was measured without the incidence of light.

Due to the difficulty of finding a thermometer that can withstand high temperatures, to measure the temperature of the concrete block we used a thermometer that measures the emission of infrared rays from the body being analysed (model gm1150a). As each substance has a certain emissivity constant, the thermometer was set to operate at a rate of 0.94, which corresponds to the thermal emissivity of concrete. Appendix E lists the emissivity of some surfaces according to the thermometer's instruction manual.

The local ambient temperature was obtained using a HOMIS model 511A Multi-Function Meter. Figure 56 shows the concentration of the spotlight on the concrete block, where the thermometer was operated 500 mm from the spotlight, and Figure 57 shows the equipment used to take the temperature measurements.

Figure 56 - Concentration of the spotlight on the concrete block
Source: CERQUEIRA; SANTOS, 2013.

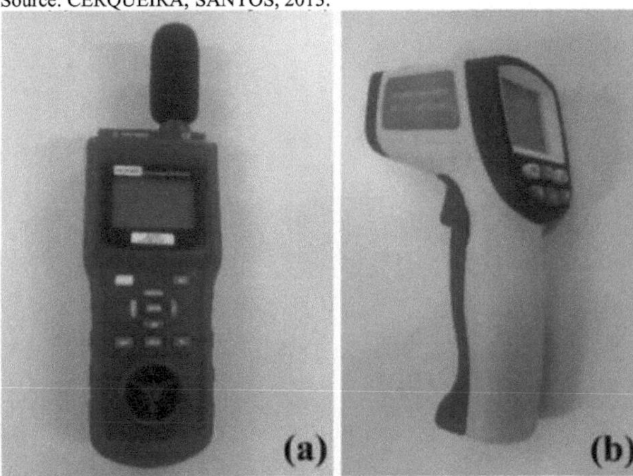

Figure 57 - Equipment used (a)Multi-function meter; (b)Infrared thermometer
Source: CERQUEIRA; SANTOS, 2013.

During the evaluations on the 15th, the weather remained clear with little cloudiness, an average ambient temperature of 32.35°C, and high solar incidence (Graph 1). Table 13 shows the relationship between reflector focal point temperature, ambient temperature and solar incidence on 15 April.

Table 13 - Temperature and solar incidence values (15 April 2013)

Local Time	Focus Temperature (°C)	Ambient Temperature (°C)	Solar Radiation Intensity (W/m $)^2$
10:00	411,1	34	659.2
10:30	540,7	33,7	-
11:00	505,3	33,2	1471
11:30	638,6	32,4	-
12:00	651,3	32,3	2379
12:30	654,8	32	-
13:00	514,8	32,5	3067
13:30	518,1	31,6	-
14:00	509,6	31,4	3309
14:30	235	31,1	-
15:00	-	-	3437

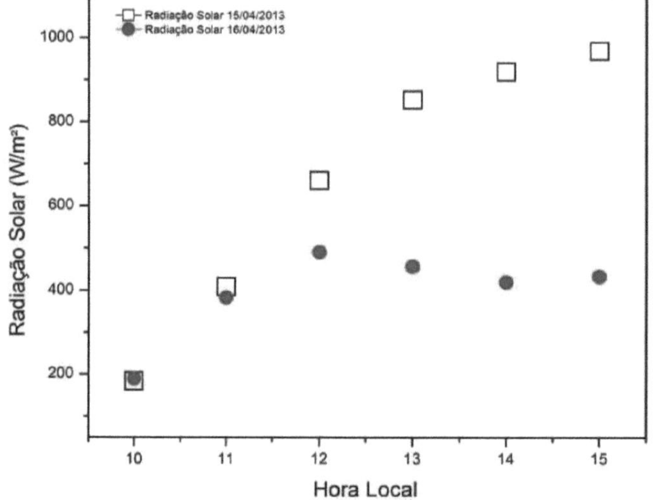

Graph 1 - Solar incidence data (15th and 16th April 2013) (INMET)

Day 15 saw the best performance from the concentrator, reaching temperatures in excess of 400°C in all evaluations up to 2pm, and even exceeding 600°C between 11.30am and 12.30pm.

At the end of the day, there was a high accumulation of heat on the concrete focus, where a temperature of 235°C was recorded. Graph 2 shows the ambient temperature and the temperature of the focus on the concrete block, while Graph 3 shows the solar incidence data and the temperature of the focus, both on 15 April.

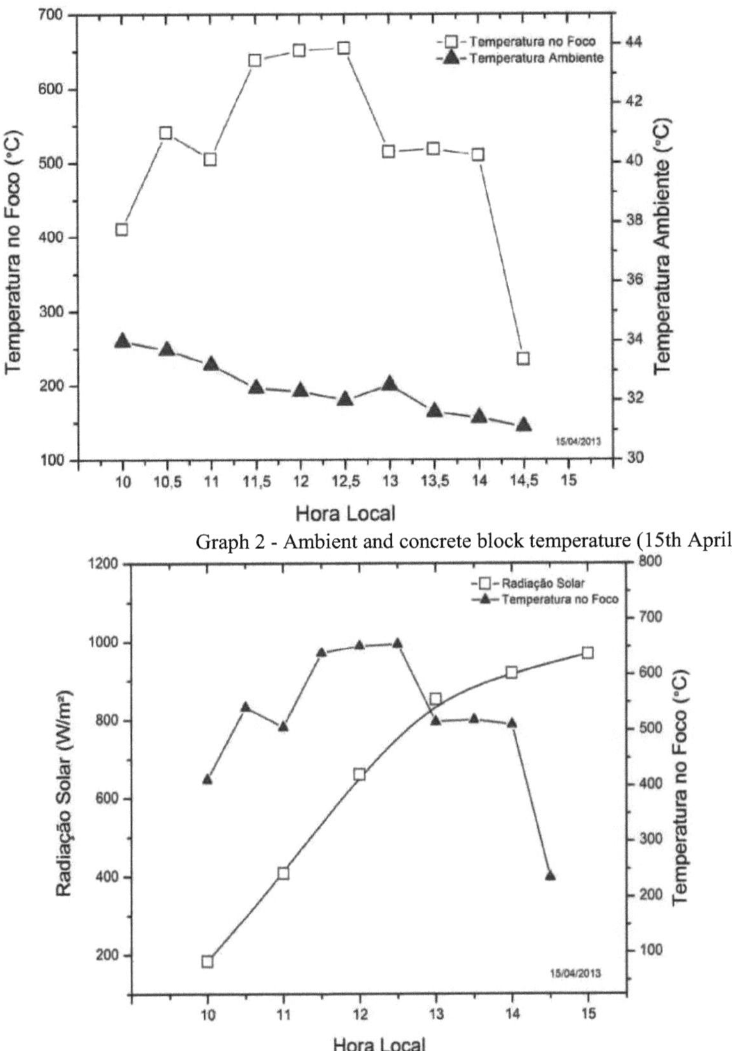

Graph 2 - Ambient and concrete block temperature (15th April)

Graph 3 - Temperature at the focus and solar incidence values (15 April)

On the second day of assessment (16 April), the weather conditions changed substantially. Although it didn't rain, the day was completely cloudy and muggy, with an average ambient temperature of 33.65°C and little direct sunlight (Graph 4). Table 14 shows the values recorded for the focus temperature, ambient temperature and solar incidence on this day.

Table 14 - Temperature and solar incidence values (16 April 2013)

Local Time	Focus Temperature (°C)	Ambient Temperature (°C)	Solar Radiation Intensity (W/m^2)
10:00	74	32	682,4
10:30	94,1	33,9	-
11:00	172,4	36,3	1378
11:30	337,6	36,3	-
12:00	132,6	33,8	1767
12:30	83,2	33,4	-
13:00	71,6	33,8	1644
13:30	209	33,5	-
14:00	84,4	32,5	1509
14:30	69,1	31,8	-
15:00	-	-	1559

The temperatures assessed on day 16 on the concrete block were relatively low, exceeding 200°C only at two points (11.30am and 1.30pm), when less dense clouds passed between the sun and the concentrator. Graph 4 and Graph 5 make it easier to understand the values measured during this test day.

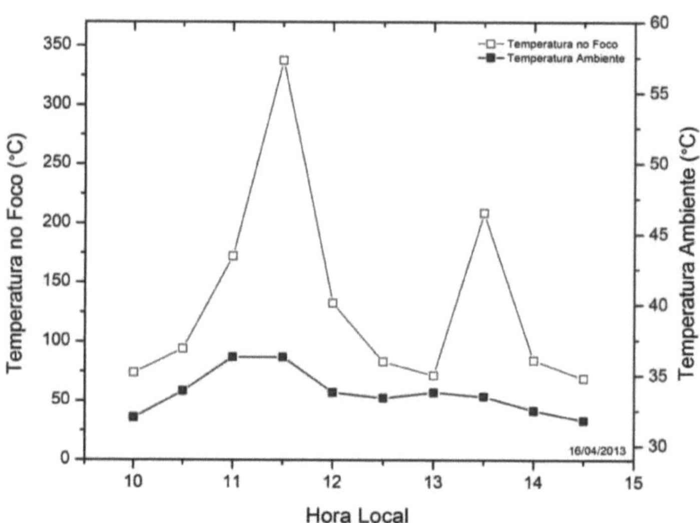

Graph 4 - Ambient and concrete block temperature (16th April)

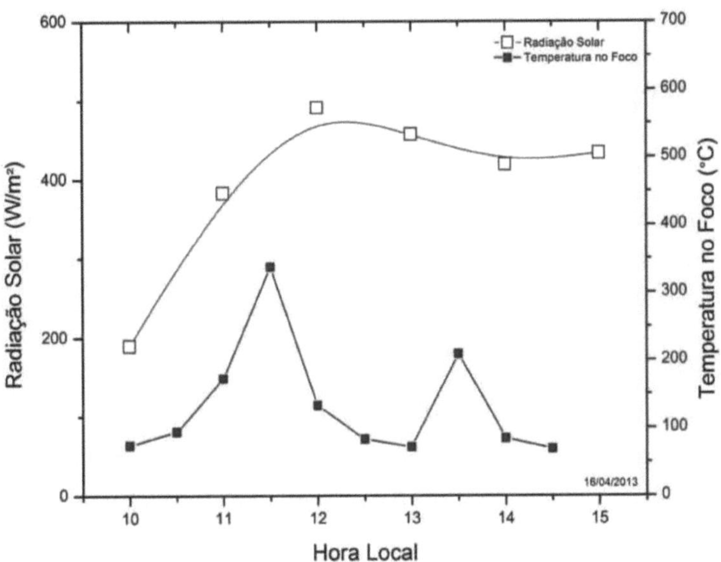

Graph 5 - Temperature at the focus and solar incidence values (16th April)

Although the progression of solar incidence on the 15th (Graph 3) was similar to that on the 16th (Graph 5) during the interval from 10am to 11am, it can be seen that the focus temperature values were quite different between the two days. Observing the difference in climate between the two days at the test site suggests that the data collected by INMET on the 15th was affected by cloudiness.

Analysing the results, it can be seen that the efficiency of the solar concentrator is influenced by the degree of solar incidence and the climatic conditions at the time of its application; and that the ambient temperature is not a determining factor for its application, since day 16 was warmer than day 15.

It can be concluded that the most favourable days for using the solar concentrator are those with clear skies and little or no cloud cover.

CHAPTER 6

FINAL CONSIDERATIONS

As solar energy is an inexhaustible and free energy, the technologies related to it are promising for a cleaner energy scenario. This work has demonstrated the applicability and ease of use of the Fixed Focus Parabolic Solar Concentrator for the cooking process, as well as its high energy potential.

The materials to build the concentrator and the cooker were easily obtained, as Wolfgang Scheffler (2006b) surmised. Building the structure of this solar concentrator proved to be relatively simple. Except for the reflector panel, which was moderately difficult, requiring more attention during construction, but nothing so difficult as to make it impossible to reproduce. As of the end of this work, the solar tracker has not been installed on the concentrator, making it a topic for future work.

With the results evaluated, it can be concluded that this Scheffler Solar Concentrator has great potential for use in community kitchens in rural areas in Brazil, especially in regions that still use firewood for cooking.

Extending this technology to remote communities, especially in the northeastern region, can be seen as a sustainable practice. As this region has favourable climatic conditions, the application of fixed-focus solar concentrators would represent a reduction in consumption and dependence on firewood in the most isolated communities, thus protecting the caatinga and cerrado biomes. It could also contribute to generating local income.

However, given the high temperatures the concentrator can reach and assessing the economic viability of building it, its use can be favourable in economic activities such as inns, restaurants and laundries. Concentrated energy can also be used to heat electrostatic painting ovens, flour mills, greenhouses, autoclaves, industrial furnaces and even to generate steam as a driving force for electric generators.

Despite the benefits and applications of the Fixed Focus Parabolic Solar

Concentrator, this technology is currently not very widespread in Brazil. It is hoped that the production of this work will make the Scheffler solar concentrator more widespread in Brazil, and that it will also help in the construction of other prototypes of this model.

CHAPTER 7

REFERENCES

ACHÃO, Carla da Costa Lopes. Analysing the Structure of Energy Consumption by the Brazilian Residential Sector. 2003. Dissertation (Master of Science in Energy Planning) - Federal University of Rio de Janeiro. Rio de Janeiro, 2003.

ALMEIDA, Iaponan Cardins de Sousa; et al. Effects and Reflections of the Caatinga as an Energy Resource in Junco do Seridó - PB. 2008. Available at: <http://www.inicepg.univap.br/cd/INIC_2008/anais/arquivosINIC/INIC1042_01_O.pdf>. Accessed on: 08 July 2013

ANEEL. National Electricity Agency. Solar Energy. Atlas of Electricity in Brazil. 2. ed. Brasília: ANEEL, 2005. chap. 3. Available at: <http://www.aneel.gov.br/aplicacoes/Atlas/download.htm>. Accessed on: 24 October 2012.

ANUÁRIO ESTATÍSTICO DO ESTADO DE ALAGOAS - ANO 2010, n. 17. Maceió: Secretaria do Estado de Alagoas, 2011.

BARBOSA, Ricardo. Passive Solar Heating. 2008. Available at: <http://planetacad.com/presentationlayer/Estudo_01.aspx?id=18&canal_orde m=0403>. Accessed on: 18 Dec. 2012.

BAUMAN, Amy. Natural Resources. Barueri, SP: Girassol, 2008 (Planet Earth).

BENEDUCE, Fábio Cezar Aidar. The Energy Society and the Environment. 2. ed. Aquiraz, CE: Iteva, 2000.

BRAZIL. Ministry of Agriculture, Livestock and Supply. National Meteorological Institute. Available at: <http://www.inmet.gov.br/sonabra/pg_dspDadosCodigo.php?QTMwMw==>. Accessed on: 26 April 2013.

CAMPBELL, Stu. Build Your Solar Heater. Portugal: Publicações Europa-América, 1978.

COENDERS, A. Química Culinaria. Acribia, Zaragoza. 1996. 290p.

DEGANUTTI, Roberto, PALHACI, Maria do Carmo Jampaulo Plácido, ROSSI, Marco et al. Rural biodigesters: Indian, Chinese and batch models. In: ENCONTRO

DE ENERGIA NO MEIO RURAL, 4., 2002, Campinas.

DELANEY, David. Notes on Scheffler: Community Kitchens. 2009. Available at: <http://davidmdelaney.com/scheffler-precis/scheffler- precis.html>. Accessed on: 21 January 2013.

DELERUE, Alberto. The Solar System. Rio de Janeiro: Ediouro, 2002.

DIB, Erick Alfred. Design and Construction of a Fixed Focus Solar Concentrator used to heat an oven. 2009. Dissertation (Master's Degree in Process Engineering) - Tiradentes University. Aracaju, 2009.

DUTRA, Ricardo Marques. The technical and economic viability of wind energy in view of the new regulatory framework for the Brazilian electricity sector. 2001. Dissertation (Sciences in Energy Planning) - Federal University of Rio de Janeiro. Rio de Janeiro, 2001.

BARROS FILHO, José Roberto Galdino de; CRUZ, Rennisy Rodrigues. The Environmental Movement in Alagoas: origins, theoretical influences, reflections and perspectives. 2011. Final Coursework (Undergraduate Degree in Environmental Management) - Federal Institute of Alagoas. Marechal Deodoro, 2011.

GOLDEMBERG, José; LUCON, Oswaldo. Energy and the Environment in Brazil. ESTUDOS AVANÇADOS 21 (59). 2007.

GOLDEMBERG, José; VILLANUEVA, Luz Dondera. Energy, Environment and Development. 2 ed. São Paulo: Editora da Universidade de São Paulo, 2003.

GOMEZ, Manu; KERN, Maren. Construction Manual 2.7m^2 Scheffler: Reflector Solar Cooker.2010. Available at:<http://www.solarebruecke.org/Bauanleitungen/2,7%20qm%20manual%2 0juli%202010.pdf>. Accessed on: 17 Nov. 2010.

GROUP 07. Solar Energy. Association of War College Graduates, Cycle of Studies on Security and Development, 1979.

HINRICHS, A. R.; KLEINBACH, M. Energia e Meio Ambiente. Translation: Flávio Maron and Leonardo Freire de Mello. Original title: Energy: its use and theenvironment. São Paulo: Pioneira Thomson Learning, 2003.

KREITH, Frank. Principles of Heat Transmission. Translation: Eitaro Yamane, Otávio de Mattos Silvares, Virgílio Rodrigues Lopes de Oliveira. São Paulo; Edgard Blucher, 1977.

LÓPEZ, Juan Carlos Flores; SILVA, Márcio Lopes da; SOUZA, Agostinho Lopes.

Residential Firewood Consumption in Cachoeira de Santa Cruz, Viçosa-MG, Brazil. Revista Árvore, v.24, n° 4, 2000. Available at: <http://books.google.com.br/books?hl=pt-BR&lr=&id=YniaAAAAIAAJ&oi=fnd&pg=PA423&dq=Consumo+residencial+d e+lenha+em+Cachoeira+de+Santa+Cruz+vi%C3%A7osa&ots=w3tEYnGMP6 &sig=J6fkVy4T_h9oMeq5Uuw5x1ziY7k#v=onepage&q&f=false> Accessed on: 10 Jul. 2013.

LUIZ, Adir M. Recursos energéticos. In: How to Harness Solar Energy. São Paulo: Edgard BlucherLtda.,1985. p 1-21.

MARCATTO, Celso. Environmental education: concept and principles. State Environmental Foundation - FEAM. 1 ed. Belo Horizonte: Editora Sigma 2002.FEAM,2002.

MARIANO, Jacqueline B. Environmental Impacts of Oil Refining. Dissertation (Master of Science in Energy Planning)-
Federal University of Rio de Janeiro (PPE-COPPE/UFRJ), 2001.

MARTINS, F. R.; GUARNIERI, R. A.; PEREIRA, E. B. O Aproveitamento da Energia Eólica. Revista Brasileira de Ensino de Física, v. 30, n. 1, 1304, 2008. Available at: <http://www.sbfisica.org.br/rbef/pdf/301304.pdf>.
Accessed on: 18 October 2012.

ORNELLAS, A. J. Energy from Ancient Times to the Present Day. Maceió: Edufal, 2006. Available at: <http://ebookbrowse.com/a-energia-dos- tempos-antigos-a aos-dias-atuais-pdf-d61912980>. Accessed on: 22 January 2013.

PASSOS, Luigi A. A. et al. The Use of the Scheffler Solar Reflector in the Cooking and Processing of Food: a Sustainable Alternative for the
Northeast Brazil. Available at:

<http://www.aetec.org.br/conferencia_internacional/trab21.htm>. Accessed on: 14 Dec. 2012.

REIS, Lineu Bélico dos; FADIGAS, Eliane A. Amaral; CARVALHO, Claúdio Elias. Energy, Natural Resources and the Practice of Sustainable Development. Barueri, SP: Manole, 2005.

RIBEIRO, Viviane Wallen Silva de Moura; BASSANI, Christina. The Question of Hydroelectric Power as an Essential Energy Source in the Current Sustainability Model: the Case of Belo Monte. CONGRESSO NACIONAL DE EXCELÊNCIA EM GESTÃO, 7, 2011, Rio de Janeiro. Electronic Proceedings.
Available
at:<http://www.excelenciaemgestao.org/Portals/2/documents/cneg7/anais/T

11_0355_1508.pdf>. Accessed on: 17 October 2012.

SAMPAIO,J. L; CALÇADA, C. S. Física: volume único. 2. ed. São Paulo: Atual, 2005 (High School).

SANGA, Godfrey Alois. Impact assessment of clean technologies and substitution of cooking fuels in urban households in Tanzania. 2004. Dissertation (Master's Degree in Energy Systems Planning) - State University of Campinas. Campinas, 2004.

SCHEFFLER, W. Development of a Solar Crematorium.SCIS INTERNATIONAL SOLAR COOKER CONFERENCE, 21, 2006.Granada-Spain, 2006.
Available at:
<http://www.solarebruecke.org/infoartikel/Papers_%20from_SCI_Conference _2006/22_wolfgang_scheffler.pdf>. Accessed on: 4 January 2013.

______ . Introduction to the Revolutionary Design of Scheffler Reflectors, 2006.Available at:
<http://images3.wikia.nocookie.net/__cb20080309213912/solarcooking/imag es/e/e1/Granada06a_Wolfgang_Scheffler.pdf>. Accessed on: 19 Nov. 2012.

______ . The Scheffler Reflector. Available at: <http://www.solare-bruecke.org/English/scheffler_e-Dateien/scheffler_e.htm>. Accessed on: 14 Nov. 2012.

TERCIOTE, Ricardo. Wind Energy and the Environment. In: ENCONTRO DE ENERGIA NO MEIO RURAL, 4, 2002, Campinas: UNICAMP, 2002.
Available at:

<http://www.feagri.unicamp.br/energia/agre2002/pdf/0085.pdf>. Accessed on: 23 October 2012.

TERRA FUNDACIÓN. A Solar Reflector Scheffler Unique in the Country. 2006. Available at: <http://www.ecoterra.org/articulos57es.html>. Accessed on: 22 January 2013.

TIRADENTES, Átalo Antônio Rodrigues. Using Solar Energy to Generate Electricity and Heat Water. 2006. Monograph (Specialisation in Alternative Energy Sources) - Federal University of Lavras, Minas Gerais, 2006.

UGUCIONE, Cássia; MACHADO, Cristine M. D.; CARDOSO, Arnaldo A. Evaluation of NO2 in the Atmosphere of Outdoor and Indoor Environments in the City of Araraquara, São Paulo. Química Nova, Vol. 32, No. 7, p. 18291833. 2009. Available at:< http://www.scielo.br/pdf/qn/v32n7/27.pdf>. Accessed on: 29 January 2013.

VALE, Ailton Teixeira do; RESENDE, Raquel. Estimating Residential Firewood Consumption in a Small Rural Community in the Municipality of São João D'Aliança, GO. Revista Ciência Florestal, v. 13, n. 2, 2003. Available at: <http://cascavel.ufsm.br/revistas/ojs-2.2.2/index.php/cienciaflorestal/article/view/1752/1021> Accessed on: 10 July 2013.

WACHBERGER, Michael; WACHBERGER, Hedy. Building with the Sun: Using Passive Solar Energy. Madrid: G. Gili, 1984.

ANNEX A
PLAN OF THE ROTATION BRACKET TEMPLATE

The illustration below helped in the construction of the Rotation Support Template, showing the locations of each part that makes it up.

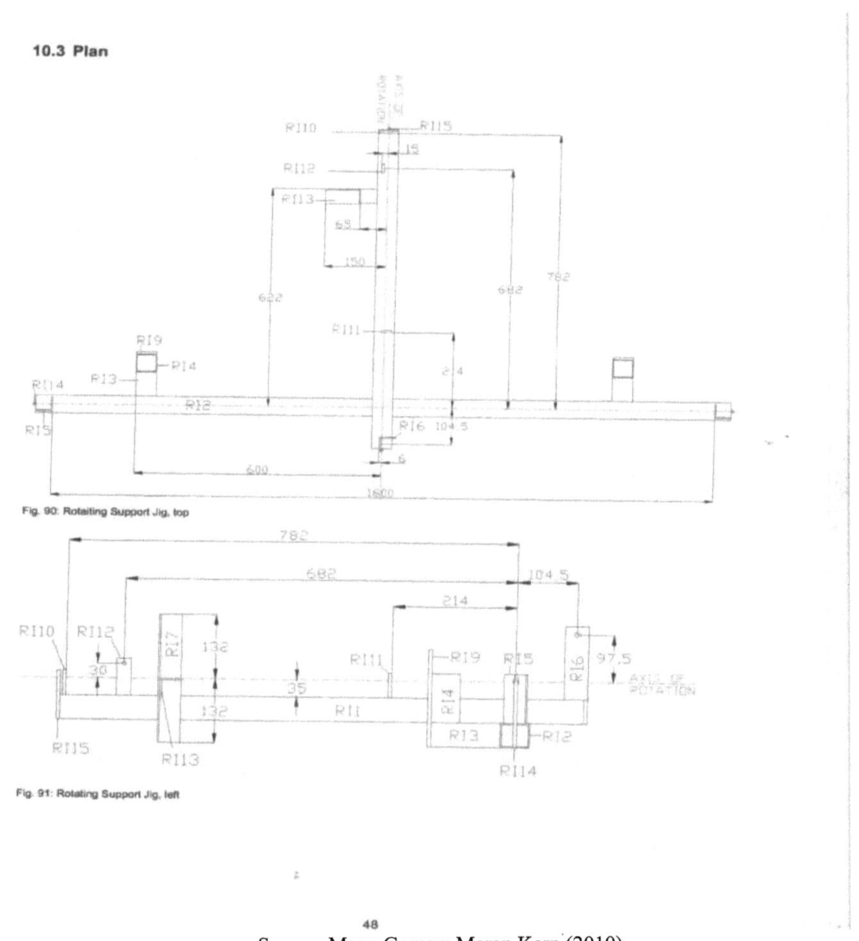

48
Source: Manu Gomez; Maren Kern (2010)

81

This illustration shows the top view of the Rotation Support plan, showing the position of each part that makes it up.

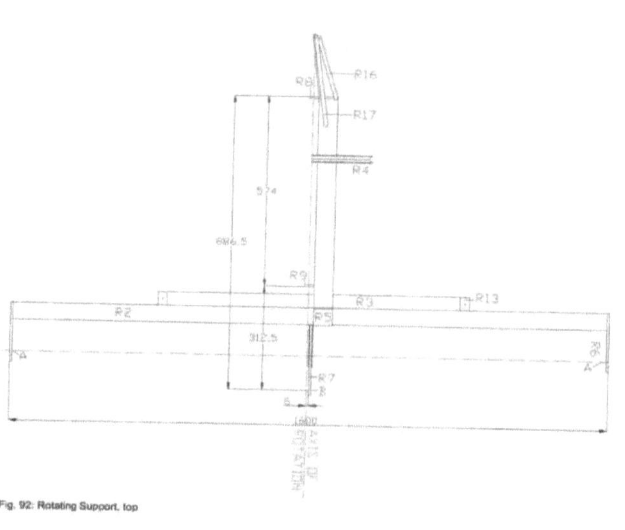

Fig. 92: Rotating Support, top

Source: Manu Gomez; Maren Kern (2010)

PLAN OF THE ROTATION BRACKET (SEEN FROM THE SIDE)

This illustration shows the right side view of the Rotation Support plan, showing the layout of each part that makes it up.

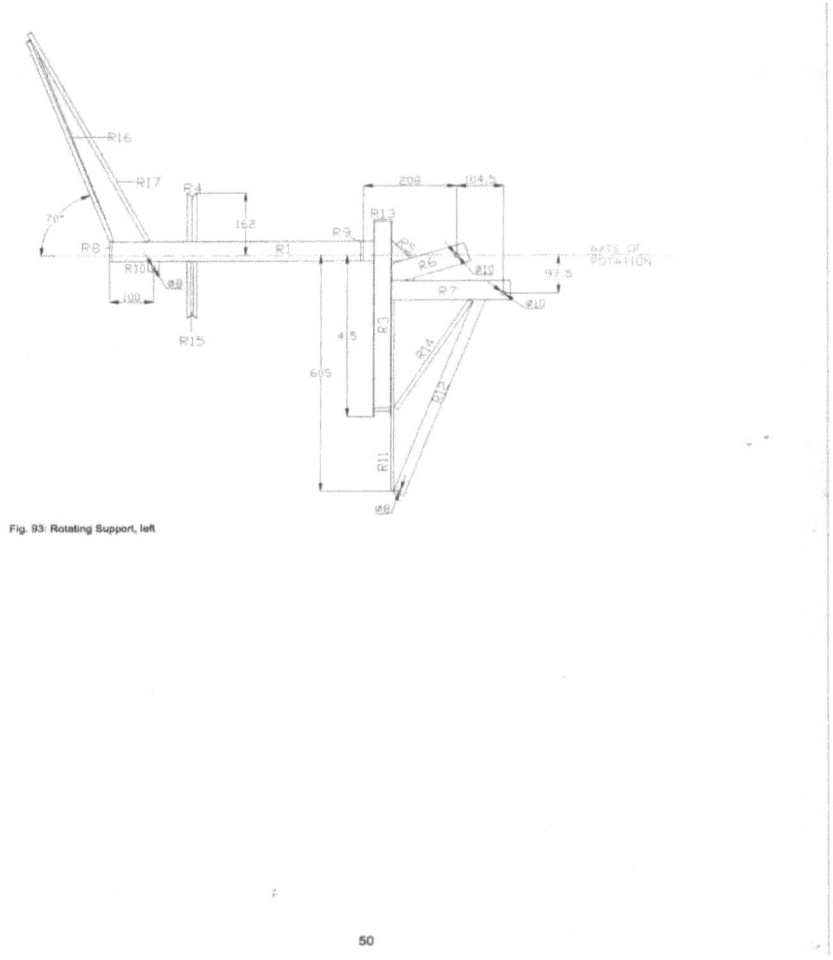

Fig. 93: Rotating Support, left

Source: Manu Gomez; Maren Kern (2010)

83

ANNEX D
PLAN OF THE ROTATION SUPPORT (SEEN FROM THE FRONT)

This illustration shows the front view of the Rotation Support plan, identifying the location of each part that makes it up.

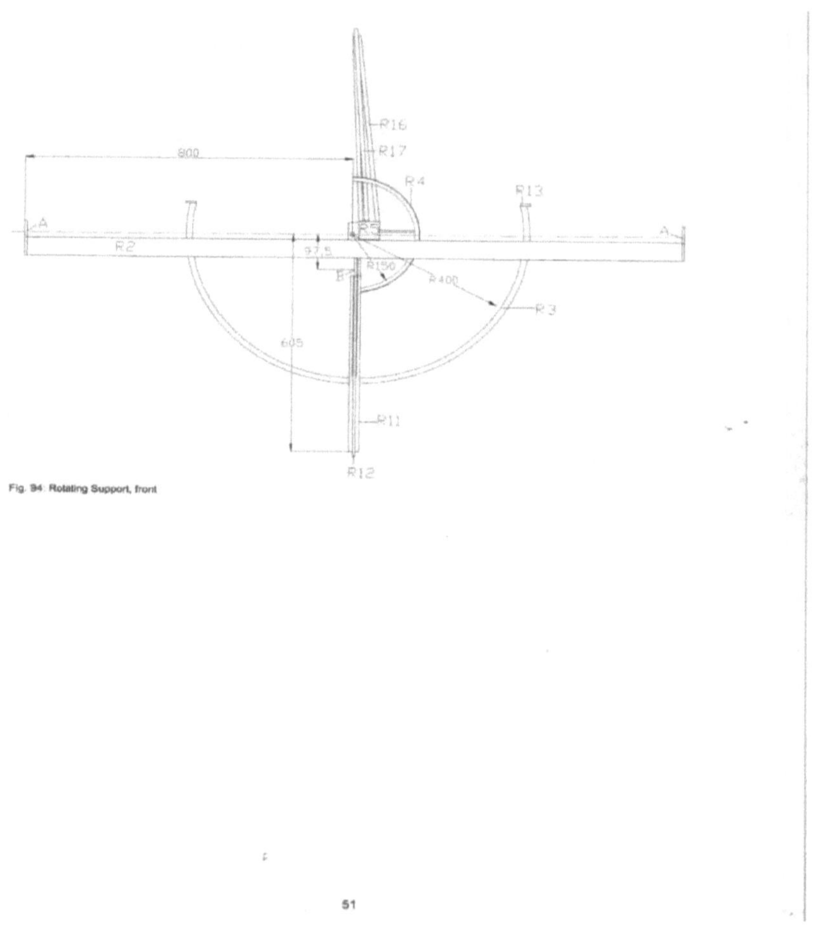

Fig. 94 Rotating Support, front

Source: Manu Gomez; Maren Kern (2010)

84

ANNEX E
LIST OF EMISSIVITY OF SOME SURFACES

The annex below helps to identify the emissivity of the surface being analysed, so that the device can be calibrated with the emissivity corresponding to the body being assessed.

The emissivity of some surfaces is listed below:

Substance	Emissivity
Asphalt	0.9 a 0.98
Concrete	0.94
Cement	0.96
Sand	0.90
Earth	0.92 a 0.96
Water	0.92 a 0.96
Ice	0.96 a 0.98
Snow	0.83
Glass	0.90 a 0.95
Ceramics	0.90 a 0.94
Marble	0.94
Plaster	0.80 a 0.90
Brick (red)	0.93 a 0.96
Cloth (black)	0.98
Human Skin	0.98
Foam	0.75 a 0.80
Charcoal (powder)	0.96
Varnish	0.80 a 0.95
Varnish (matt)	0.97
Rubber (black)	0.94
Plastic	0.85 a 0.95
Madeira	0.90
Paper	0.70 a 0.94
Chromium oxide	0.81
Copper oxide	0.78
Iron oxide	0.78 a 0.82
Fabrics	0.90

More
Books!

info@omniscriptum.com
www.omniscriptum.com
OMNIScriptum